12th International Conference on Applied Mathematical Programming and Modelling (APMOD 2016)

ITM Web of Conferences Volume 14 (2017)

Brno, Czech Republic
8 - 10 June 2016

Editor:

Ronald Hochreiter

ISBN: 978-1-5108-5192-4

Printed from e-media with permission by:

Curran Associates, Inc.
57 Morehouse Lane
Red Hook, NY 12571

Some format issues inherent in the e-media version may also appear in this print version.

Printed by Curran Associates, Inc. (2017)

For additional information, please contact EDP Sciences – Web of Conferences
at the address below.

EDP Sciences – Web of Conferences
17, Avenue du Hoggar
Parc d'Activité de Courtabœuf
BP 112
F-91944 Les Ulis Cedex A
France

Phone: +33 (0) 1 69 18 75 75
Fax: +33 (0) 1 69 28 84 91

contact-edps@webofconferences.org

Additional copies of this publication are available from:

Curran Associates, Inc.
57 Morehouse Lane
Red Hook, NY 12571 USA
Phone: 845-758-0400
Fax: 845-758-2633
Email: curran@proceedings.com
Web: www.proceedings.com

TABLE OF CONTENTS

APMOD (APplied mathematical programming and MODelling) is a series of conferences which was started in 1991 due to an initiative of Gautam Mitra at Brunel University/West London. This was then followed by events which were held in Budapest (1993), Brunel (1995), Limassol/Cyprus (1998), Brunel (2000), Varenna/Italy (2002), Brunel (2004), Madrid (2006), Bratislava (2008), Paderborn (2012), and Warwick (2014).

The 12th International Conference on Applied Mathematical Programming and Modelling (APMOD 2016) will be held in Brno, Czech Republic in June 8-10, 2016.

The conference accepts contributions from modeling, solving methods and software as well as applications of optimization. Furthermore, we are welcoming contributions from the field of data science as well as business analytics especially in relation to optimization.

Relevant application areas include, but are not restricted to finance, transportation, supply chain planning and management, revenue management, energy and infrastructure.

The models may be deterministic or stochastic - aspects considering robustness or disruption management are welcome. The methodological focus of the previous conferences has been on mathematical programming including stochastic programming, and more recently also heuristics, metaheuristics and simulation.

edp sciences

ITM Web of Conferences

Statement of Peer review

In submitting conference proceedings to ITM Web of Conferences, I certify to the Publisher that I adhere to the **Policy on Publishing Integrity** of the journal in order to safeguard good scientific practice in publishing.

1. All articles have been subjected to peer review administered by the proceedings editors.
2. Reviews have been conducted by expert referees, who have been requested to provide unbiased and constructive comments aimed, whenever possible, at improving the work.
3. Proceedings editors have taken all reasonable steps to ensure the quality of the materials they publish and their decision to accept or reject a paper for publication has been based only on the merits of the work and the relevance to the journal.

Title, date and place of the conference

The 12th International Conference on Applied Mathematical Programming and Modelling
APMOD 2016
June 8-10, 2016
Brno, Czech Republic

Proceedings editor(s):

PD Dr. Ronald Hochreiter
WU Vienna University of Economics and Business
1020 Vienna
Austria

Date and editor's signature

Vienna, March 13th, 2017

ITM Web of Conferences **14**, 00001 (2017) DOI: 10.1051/itmconf/20171400001
APMOD 2016

Applied Mathematical Programming and Modelling 2016

Ronald Hochreiter[1,*]

[1] *WU Vienna University of Economics and Business, Welthandelsplatz 1, 1020 Vienna, Austria.*

Abstract. The 12th International Conference on Applied Mathematical Programming and Modelling (APMOD 2016) was held in Brno, Czech Republic in June 8-10, 2016. In this volume eight research papers are collected which depict the broad range of topics the conference series covers.

1 Introduction

The 12th International Conference on Applied Mathematical Programming and Modelling (APMOD 2016) was held in Brno, Czech Republic in June 8-10, 2016. APMOD (APplied mathematical programming and MODelling) is a series of conferences which was started in 1991 due to an initiative of Gautam Mitra at Brunel University/West London. This initial conference was then followed by events which were held in Budapest (1993), Brunel (1995), Limassol/Cyprus (1998), Brunel (2000), Varenna/Italy (2002), Brunel (2004), Madrid (2006), Bratislava (2008), Paderborn (2012), and Warwick (2014).

The conference accepts contributions from modeling, solving methods and software as well as applications of optimization. Furthermore, we are welcoming contributions from the field of data science as well as business analytics especially in relation to optimization. Relevant application areas include, but are not restricted to finance, transportation, supply chain planning and management, revenue management, energy and infrastructure. The models may be deterministic or stochastic - aspects considering robustness or disruption management are welcome. The methodological focus of the previous conferences has been on mathematical programming including stochastic programming, and more recently also heuristics, metaheuristics and simulation.

In this volume eight research papers are collected which depict the broad range of topics the conference series covers.

In *Explaining the Results of an Optimization-Based Decision Support System – A Machine Learning Approach* by Michael Morin, Rallou Thomopoulos, Irene Abi-Zeid, Maxime Leger, Francois Grondin, and Martin Pleau the authors present their work and explain the results of a commercial software used for real-time decision support for the flow management of a combined wastewater network. The tool is deployed in many major cities and is used on a daily basis and the results show that their approach is very promising according to reactions from analysts and potential users.

The next paper *Multicommodity Flow Problems with Commodity Compatibility Relations* by Zhiyuan Lin and Raymond Kwan is concerned with a class of Multicommodity Flow Problems with

* e-mail: ronald.hochreiter@wu.ac.at

Commodity Compatibility Relations (MCFP-CCR), in which compatibility relations among commodities used at each node are required. This class of problems has application in the Train Unit Scheduling Problem (TUSP), where train units of different traction types may not be coupled with each other to serve the same train trip.

Named Entity Recognition: Resource Constrained Maximum Path by Luigi Di Puglia Pugliese, Elisabetta Fersini, Francesca Guerriero, and Enza Messina considers the open research field of Information Extraction (IE) which relates to Named Entity Recognition (NER). In their work they investigate one of the most promising model for NER based on Conditional Random Fields (CRFs). CRFs are enhanced in a two stages approach to include logic rules in the decision process that can be either extracted from data or defined by domain experts.

In their paper *MIP models for the irregular strip packing problem: new symmetry breaking constraints* Marcos Okamura Rodrigues, Luiz Henrique Cherri, and Leandro Resende Mundim consider the irregular strip packing problem which minimizes the length used to cut a set of pieces from a board with fixed width. A set of new symmetry breaking constraints are proposed to improve previously proposed models. Their results show the proposed formulation is better than the previous one for most instances.

Jan Navratil presents *Aerodynamic shape optimization of an airliner elastic wing* and compares with optimization of the same wing assuming rigid structure. A gradient-based optimization approach is applied in connection with adjoint method used for calculation of gradients of aerodynamic forces. The initial wing shape is parameterized by Free Form Deformation technique. The aerodynamics of the elastic wing is calculated using the fluid-structure interaction method which couples computational fluid dynamic solver with linear structural solver.

Emergency logistics is acquiring a crucial role aiming to provide adequate care and support for people affected by catastrophic situations; dealing with a large number of stakeholders, scarcity of resources and high stakes. Considering the importance of balancing resources from different participants to improve operations, Oscar Rodriguez-Espindola, Pavel Albores, and Christopher Brewster introduce a bi-criteria response optimisation model aiming to maximise the service provided to disaster victims while making efficient use of resource in their paper *Resource management for flood response in Mexico*.

Terminal Control Area Aircraft Scheduling and Trajectory Optimization Approaches by Marcella Sama, Konstantin Palagachev, Andrea D'Ariano, Matthias Gerdts, and Dario Pacciarelli is supporting aviation authorities with optimization methods to better use the available infrastructure and better manage aircraft movements. Their paper deals with the real-time scheduling of take-off and landing aircraft at a busy terminal control area and with the optimization of aircraft trajectories during the landing procedures.

Finally, Christoph Waldhauser and Ronald Hochreiter combine the knowledge of Operations Research and Data Mining and apply this to the field of Political Science in their paper *Shaking the trees: Abilities and Capabilities of Regression and Decision Trees for Political Science*. They compare the established method of analyzing roll call data using W-NOMINATE scores to a data-driven supervised machine learning method: Regression and Decision Trees (RDTs). Two scenarios are defined, one pertaining to an analytical goal, the other being aimed at predicting unknown voting behavior.

ITM Web of Conferences **14**, 00002 (2017) DOI: 10.1051/itmconf/20171400002
APMOD 2016

Explaining the Results of an Optimization-Based Decision Support System – A Machine Learning Approach

Michael Morin[2,*], *Rallou* Thomopoulos[3], *Irène* Abi-Zeid[1,**], *Maxime* Léger[1], *François* Grondin[4], and *Martin* Pleau[4]

[1] *Department of Operations and Decision Systems, Université Laval, Québec, Canada*
[2] *Department of Computer Science and Software Engineering, Université Laval, Québec, Canada*
[3] *IATE Joint Research Unit, INRA Montpellier, France*
[4] *Tetra Tech QI inc., Québec, Canada*

Abstract. In this paper, we present work conducted in order to explain the results of a commercial software used for real-time decision support for the flow management of a combined wastewater network. This tool is deployed in many major cities and is used on a daily basis. We apply decision trees to build rules for classifying and interpreting the solutions of the optimization model. Our main goal is to build a classifier that would help a user understand why a proposed solution is good and why other solutions are worse. We demonstrate the feasibility of the approach to our industrial application by generating a large dataset of feasible solutions and classifying them as satisfactory or unsatisfactory based on whether the objective function is a certain percentage higher than the optimal (minimum) objective. We evaluate the performance of the learned classifier on unseen examples. Our results show that our approach is very promising according to reactions from analysts and potential users.

1 Introduction

Decision support systems (DSS) and in particular optimization systems are more and more common in engineering, business, and management environments. Although these systems may be very sophisticated and powerful in terms of solving complex mathematical models and providing optimal recommendations, more often than not, the users of these systems are skeptical of the results presented. This leads to a resistance and a mistrust with regards to these technologies. More than ever, it is becoming important to explain the solutions of the optimization systems in terms that users and operators can understand, in order to be comfortable in following or challenging the recommendations of these systems. In this paper, we present work conducted in order to explain the results of a commercial software used for real-time decision support for the flow management of a combined wastewater network. This tool is deployed in many major cities and is used on a daily basis. Over the years, our industrial partner has identified the need for explanations of the optimization results which prompted us to define a collaborative project.

*e-mail: Michael.Morin.3@ulaval.ca
**e-mail: Irene.Abi-Zeid@osd.ulaval.ca

Du *et al.* [1] distinguish between "backward" and "forward" explanations. Backward explanations are past-oriented. They seek reasons in the past (actions, events) that led to the observed situation. Forward explanations are future-oriented. They foresee the future consequences of a possible decision. In the present paper, we propose a forward approach which provides explanations for the recommendations of an optimization-based DSS. We apply a machine learning technique to build rules for classifying and interpreting the solutions of the optimization model, namely decision trees. Our main goal is to build a classifier that would help a user (not a researcher or a practitioner from a field related to optimization) to understand why a proposed solution is good and why other solutions are worse.

The paper is structured as follows: Section 2 presents a brief literature review, Section 3 includes the problem background and methodology, Section 4 contains a description of the experiment and the results. We conclude in Section 5.

2 Literature Review

Decision support systems have been around for several decades now. Interestingly, since the 90s, several studies have pointed out that end users have a low acceptance level of the solutions computed by DSSs [2–5]. These studies have explored, respectively, the importance of feedback about a decision made, the role of cognitive biases in human decision errors, the complexity of the requirements prioritization which implies an iterative decision-making process, and the importance of contextual and causal information in forecasting.

However, during the 2000s, studies about the effects of providing explanations on DSS advice acceptance were still extremely rare [6] and they have only recently begun to emerge as an important consideration [7]. In their paper, Du *et al.* [7] use two machine-learning techniques, namely rough sets and dependency networks, to mine DSS solutions. The mining results are then provided as explanations of the DSS solutions. Both techniques provide different, but not disjoint, results, which are used by the authors to classify the mining results as strong or weak explanations.

In [6], two types of explanations are distinguished: technical ones, tracing the process used to make a forecast, and managerial ones, providing a justification based on the meaning of the forecast. Both types of explanations were estimated to increase decision-makers confidence. This diagnostic is confirmed by Gonul *et al.* [8]. Based on their review and conclusions, we have built the following three tables which summarize some conclusions about the categories, the roles and the added value of explanations. These are presented respectively in Tables 1, 2 and 3.

3 Background and Methodology

The commercial DSS that we are interested in is a software for real time control and management of a combined wastewater network, provided by our industrial partner. It includes a simulator and an optimizer. The network is made-up of controlled and uncontrolled sections. Based on weather forecasts over a given time period, the simulator simulates the flow in the uncontrolled parts of the network that provide the inputs to the controlled parts of the network. The optimizer's role is to provide the complete set of control point configurations (decision variables values) that minimizes the overflow cost (objective function) for the considered period. It establishes, based on water flows in the network, a set of instructions that are sent to local stations to regulate flows using sluice gates and pumps. Constraints describe the behavior of the network and the cost function is defined to minimize overflow volumes, storage facility dewatering time, to ensure a balanced hydraulic load distribution throughout the network and to minimize the associated management costs [9]. The provided solution,

DOI: 10.1051/itmconf/20171400002

Table 1. Categories of explanations

Categories	Subcategories	Remarks
Depending on the content of explanations	Explanations on the **how**: system reasoning mechanism (information used, rules, steps)	
	Explanations on the **why**: justification of a decision highlighting the reasons behind the decision	The most efficient for the acceptance of the result provided
	Strategic explanations: overview of the whole issue and the stakes tackled by the system	Less used since their obtention is difficult
Depending on the way of providing explanations	On user demand	
	Automated	
	Intelligent: when a need for explanations is detected by the system	Complex mechanism
Depending on the format in which explanations are presented	Text-based: predefined sentences or combinations following some rules	Natural language-style sentences improve understanding and acceptance of the system
	Multimedia: graphs, pictures, animations	Costly implementation but persuasive and efficient to obtain users' confidence

Table 2. Roles of explanations

Objectives	Benefits for users
Explain a detected anomaly	Understand a recommendation
	Verify that it satisfies the expectations
	Solve a contradiction between the users and the system
Provide complementary information	Participate in solving a problem (short-term)
Facilitate learning	Perform better in future tasks (long-term)

DOI: 10.1051/itmconf/20171400002

Table 3. Added value of explanations

Effects on users	Consequences	Result
Better understanding Learning	Perform better decision making (accuracy and speed)	Accept the system as logical and thus grounded and useful
Follow the logical reasoning used by the system	Positive perception of the system (ease of use, usefulness, satisfaction, confidence)	

although optimal with respect to the objective function in the model used by the optimizer, can be modified by the end-user.

A solution fully describes, in terms of the values of the decision variables, the state of the controlled section of the network. Decision variables determine, in particular (but not exclusively), the positions of all the control gates regulating the wastewater system flow management. These can be closed or open to a certain extent.

3.1 Methodology

Given an optimal solution to an optimization problem provided by solving a mathematical program, in this case a minimization problem, we wish to enable an end-user to understand which characteristics of a solution render it more or less suitable (satisfactory) from the perspective of the decision s/he has to make. From a single objective optimization perspective, a solution that is *optimal* minimizes a predefined and accepted objective function while being feasible, *i.e.*, while respecting a set of predefined constraints. We employ the term *satisfactory* to describe a feasible solution that is a suitable decision in the context where a mathematical program is used as part of a DSS. An optimal solution is of course satisfactory. However, some satisfactory solutions might be suboptimal.

Suppose, for instance, that a small increment δ in the objective function value is tolerable. Any feasible solution x with an objective value $f(x)$ that is not worse than $f(x^*)+\delta$, *i.e.*, $f(x) \leq f(x^*)+\delta$, is said to be satisfactory. Although x is not optimal, it is a suitable decision with respect to that criterion.

We therefore need a methodology to characterize the satisfactory solution set, using simple rules, in order to allow a DSS user to understand the following:

1. why a solution (including the optimal one) is satisfactory; and

2. why a given change to an optimal solution might produce a solution that is worse, or potentially catastrophic in practice.

If a user is not satisfied with the optimal solution initially provided by the DSS for a number of possible reasons, including the fact that it does not intuitively make sense to her/him, the rules (explanations) would allow her/him to understand which decision variables can be modified (and how) before the DSS can re-optimize and return a new satisfactory solution.

We chose to use decision trees to produce (learn) such a simple set of rules to characterize the set of satisfactory solutions. This is similar to the approaches in [7], albeit in a different context and on bigger-size data sets.

 DOI: 10.1051/itmconf/20171400002

3.2 Decision trees and the classification of satisfactory solutions

Decision trees are a widely used machine learning technique. In the next subsections, we briefly present machine learning and decision trees as we applied them to the problem of characterizing the set of satisfactory solutions.

3.2.1 Machine learning

(Supervised) machine learning is the art of learning by examples [10]. A learning algorithm is provided with a training set S of examples. Each example in set S is a pair (x, y) where x is an input and y is an output. The output can be a class, a real number, or it can have a structure [11]. When $x \in \mathbb{R}^n$ and $y \in \{True, False\}$, we have a *binary classification* problem. We call the entries in x the features and y the class.

A solution to the binary classification problem is a function h, called a classifier, that takes an input $x \in \mathbb{R}^n$ and that outputs $y \in \{True, False\}$. A learning algorithm builds a classifier h using the set S, *i.e.*, known pairs (x, y) of features and class (*training phase*). Once built, the classifier can be used to compute an output class y given an unseen vector of input features x (*testing phase*). In this paper, we formulate the problem of recognizing satisfactory solutions from unsatisfactory ones as a binary classification problem.

3.2.2 Decision trees

A decision tree [12] is a classifier in the form a tree structure. Classifiers can be either *white box* or *black box* models. We chose to use decision trees since a decision tree is a white box and interpretable model made of simple decision rules. It takes x as an input, and it outputs a class y. The nodes of the tree are either decision nodes or leaves. A *decision node* corresponds to a branching rule based on the value of a single input feature. A *leaf* node is assigned a prediction class.

In the training phase, a decision tree is built in a greedy fashion by the learning algorithm. The examples of the training set S, *i.e.*, the known examples, start at the root of the tree. A first branching rule is applied at the root on the training examples. The learning algorithm chooses the branching rule that best partitions the training examples according to a heuristic. All examples x that respect the branching rules take the left path. All other examples take the right path. At the next level, a new decision node is created for each partition and two branching rules are selected according to the partitioning heuristic. Branching stops when all the examples of a partition belong to the same class. Other stopping criteria to manage the height of the tree can be used. In that case, not all the examples in a partition that reach a leaf will have the same class and the value of the leaf can be set to the majority class.

In the testing phase, an unseen example x, *i.e.*, an example for which y is unknown, follows a path from the root to a leaf. The example is tested against each decision rule. If it respects the rule, it branches left, otherwise it branches right. The predicted class is that of the reached leaf.

3.2.3 Learning to recognize satisfactory solutions

As mentioned previously, each node of a decision tree, except the leaves, is a branching rule. A path from the root to a leaf can be seen as a rule describing the satisfactory solution set. We first learn a decision tree (training phase). Later on, this decision tree is used to determine whether a solution is satisfactory or not (test phase). Furthermore, since a decision tree is interpretable, it provides a set of

DOI: 10.1051/itmconf/20171400002

rules characterizing satisfactory solutions. We next describe the process used to generate the set S of training examples that can be used for learning.

Assume that an optimal solution has been provided by the DSS and that a satisfactory criterion has been determined. We generate set S by simulating a sufficiently large number of modifications of the optimal solution. Simulating a modification consists in randomly fixing a decision variable's value. After each simulation, we evaluate the value of the solution according to the chosen criterion that discriminates between satisfactory and unsatisfactory ones. We store in S, the value of all decision variables along with the class of the newly generated solution: satisfactory (true) or not (false). Finally, we learn one or more decision trees on S.

4 Experiment

In order to demonstrate the feasibility of the approach to our industrial application, we generated a large dataset of feasible solutions and classified them as satisfactory or not based on whether the objective function is a certain percentage higher than the optimal objective. We then used a small subset of these solutions to learn a classifier. The small subset used for learning is the training set (S). We evaluated the performance of the learned classifier on the remaining unseen examples. We provide the details on the data we generated for the experiment in the next subsection.

4.1 Data preparation

In order to learn the characteristics of the satisfactory solution set, we first obtained an optimal solution to a problem instance using the optimizer of the DSS developed by our partner. The weather forecast, the simulation results, and the provided optimal solution belong to a real use-case scenario.

Starting from the optimal solution, we generated 10,000 feasible solutions by simulating a single modification of one of the decision variables by the end-user. Each solution of the dataset used for the experiment is generated as follows:

- one of the decision variables is chosen and modified randomly;
- the objective value of the solution resulting from the perturbation of that single control point is reevaluated.

The resulting dataset contains the values of all the variables of the optimization model for all the generated solutions along with their objective value. We then filter the dataset to keep only the variables that can have an explanatory significance for the end-user. The identification of these variables was performed by the domain experts. We excluded all the variables that are used as components of the objective function. We then assigned a class to each solution. For the purpose of the experiment, we assume that a solution is nonsatisfactory if and only if its objective value exceeds by 5% (or more) the optimal value.

In the next subsection, we present, the results of our methodology when used jointly with decision trees.

4.2 Decision trees and their predictive accuracy

All decision tree experiments are conducted using scikit-learn [13], a machine learning library for the Python programming language.

Figure 1 shows the learning curve we obtain on our industrial instance. Most learning algorithms for decision trees are greedy algorithms. Scikit-learn's learning algorithm for decision trees makes no

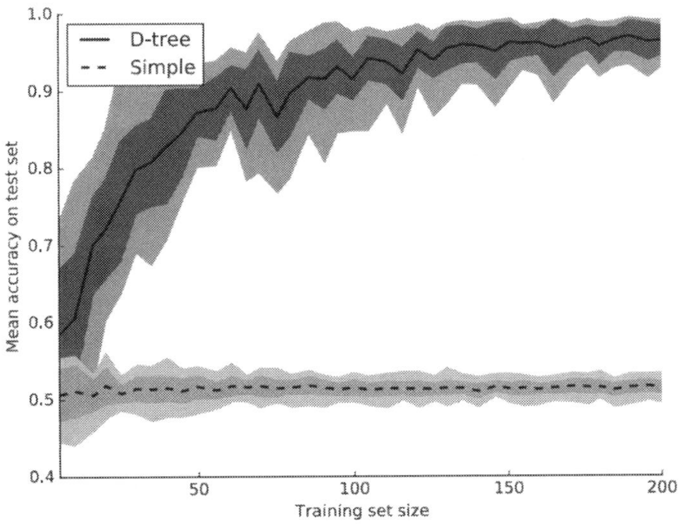

Figure 1. Learning curve: Mean accuracy of the learned decision trees (solid curve) and of the simple classifiers (dashed curve) on the test set against various training set sizes with standard deviation in blue and green shading and upper and lower bounds in light blue and light green

exception. In order to evaluate the predictive accuracy of decision trees in our context, we partitioned the dataset into a training set S and a test set. Some partitions might be more or less satisfactory for learning. We aggregated the results of 10 different partitions. We generated the curve of Figure 1 by varying the size of the subset of S used for training. For each subset of S, ranging in size from 5 to 200 training examples (for a total of 40 subset sizes), we learned 10 different trees and aggregated the results. Each tree was tested against the 9,800 remaining examples in the test set. Using such a large test set enables us to verify whether a decision tree learning algorithm can generate rules to characterize the satisfactory solutions space of our industrial problem instance by accessing a few training examples only. The solid curve plots the average predictive accuracy of the decision trees against the size of the training subsets used for learning. The blue shaded region shows the standard deviation. The light blue shaded region shows the minimal and the maximal predictive accuracy encountered for each size of subset of the training set. The dashed curve plots the average predictive accuracy achieved by the mean of simple classifiers generating their prediction using the distribution of the classes in the training set. The standard deviation region and the region of the bounds of the simple classifiers are displayed in green.

The decision rules learned by the decision trees were validated by experts from the application field. The tendency of the learning algorithm was to choose high impact control points (decision variables) in its first few decision rules.

5 Conclusion

In this study, we have proposed an approach using decision trees to explain the results of a commercial software used for real-time decision support for combined wastewater system flow management. Our solution allows analysts to identify the main criteria that influence the DSS outcome, and helps users in understanding the rationale behind the DSS's recommendations. The rationale supporting this approach consists in classifying the space of solutions according to some criterion that characterizes the performance of the solutions proposed by the DSS.

We plan to explore several directions in the next stages of this work. First, we will identify "optimal" sizes of the training set and tree depths, allowing for both a good learning quality and a reasonable computing time. Moreover, the robustness of the method will be tested based on the following factors: on the one hand, the number of variables taken into account, in particular for increased numbers of predictive variables; on the other hand, the proportion of positive examples composing the training set. Finally, we plan to experiment our approach on other kinds of learning algorithms, also able to compute interpretable classifiers, with the long-term objective of providing users with arguments expressing the learnt rules.

References

[1] G. Du, M. Richter, G. Ruhe, *An explanation oriented dialogue approach and its application to wicked planning problems*, Journal of Computing and Informatics **25**(2–3), 223 (2006)

[2] F.D. Davis, J.E. Kottemann, *Determinants of decision rule use in a production planning task*, Organizational Behavior and Human Decision Processes **63**(2), 145 (1995)

[3] B. Kleinmuntz, *Why we still use our heads instead of formulas: Toward an integrative approach*, Psychological Bulletin **107**(3), 296 (1990)

[4] L. Lehtola, S. Kujala, *Requirements prioritization challenges in practice*, in *Proc. of the 5th International Conference on Product Focused Software Process Improvement (PROFES)* (2004), pp. 497–508

[5] J.S. Lim, M. O'Connor, *Judgmental forecasting with time series and causal information*, International Journal of Forecasting **12**(1), 139 (1996)

[6] M. Lawrence, L. Davies, M. O'Connor, P. Goodwin, *Improving forecast utilization by providing explanations*, in *Proc. of the 21st International Symposium on Forecasting (ISF)* (2001)

[7] G. Du, G. Ruhe, *Two machine-learning techniques for mining solutions of the ReleasePlanner™ decision support system*, Information Sciences **259**, 474 (2014)

[8] M.S. Gönül, D. Önkal, M. Lawrence, *The effects of structural characteristics of explanations on use of a DSS*, Decision Support Systems **42**(3), 1481 (2006)

[9] M. Pleau, H. Colas, P. Lavallée, G. Pelletier, R. Bonin, *Global optimal real-time control of the Quebec urban drainage system*, Environmental Modelling & Software **20**(4), 401 (2005)

[10] C.M. Bishop, *Pattern Recognition and Machine Learning* (Springer, 2006)

[11] G. Bakĭr, T. Hofmann, B. Schoölkopf, A.J. Smola, B. Taskar, S.V.N. Vishwanathan, eds., *Predicting Structured Data* (MIT press, 2007)

[12] L. Breiman, J. Friedman, R.A. Olshen, C.J. Stone, *Classification and Regression Trees* (Chapman and Hall, 1984)

[13] F. Pedregosa, G. Varoquaux, A. Gramfort, V. Michel, B. Thirion, O. Grisel, M. Blondel, P. Prettenhofer, R. Weiss, V. Dubourg *et al.*, *Scikit-learn: Machine learning in Python*, The Journal of Machine Learning Research **12**, 2825 (2011)

ITM Web of Conferences **14**, 00003 (2017) DOI: 10.1051/itmconf/20171400003
APMOD 2016

Multicommodity Flow Problems with Commodity Compatibility Relations

Zhiyuan Lin[1],* and *Raymond* Kwan[1],**

[1] *School of Computing, University of Leeds, Leeds, United Kingdom, LS2 9PT*

Abstract.
We present a class of Multicommodity Flow Problems with Commodity Compatibility Relations (MCFP-CCR), in which compatibility relations among commodities used at each node are required. This class of problems has application in the Train Unit Scheduling Problem (TUSP) [1, 2], where train units of different traction types may not be coupled with each other to serve the same train trip. Computational complexity issues are discussed and solution methods are proposed. Computational experiments using the proposed solution methods are reported.

1 Introduction

The Multicommodity Flow Problem [3] is characterized by a set of commodities $k \in K, |K| > 1$ to be flowed through a network represented by a graph formed by a node set N and an arc set A as $G = (N, A)$. Each commodity k has a source s_k and a sink t_k. Different types of the problem may differ in features such as minimum supply requirements and/or capacity bounds shared by all commodities on arcs and/or nodes, and other side constraints.

The multicommodity flow problem can be traced back to Ford and Fulkerson [4]. Surveys on multicommmdity flow problems can be found in Hu [5], Kennington [6], Ahuja et al. [3]. In general, the continuous Multicommodity Flow Problem is solvable by a polynomial LP solver [3] while its integer version is $\mathcal{NP}$-hard [7, 8].

In a standard multicommodity flow problem, if a nodes and/or an arc is allowed to be flowed by different types of commodities, commodity flows can share the node and/or arc as long as all constraints are satisfied. In this paper, *compatibility relations* are imposed on commodities used at nodes and/or arcs where only certain subsets of them can coexist. For instance, when five commodities k_1, k_2, k_3, k_4, k_5 are allowed at node $j \in N$, commodities used at j should be either from k_1 and k_2 or from k_3, k_4 and k_5, but not otherwise. We refer to this problem as the Multicommodiy Flow Problem with Commodity Compatibility Relations (MCFP-CCR). If the flows are required to be integers, then the corresponding problem is called Integer MCFP-CCR, or IMCFP-CCR. *Families of commodities* are established, where commodities are compatible within the same family. Moreover, families will partition commodities, i.e. no commodity can be compatible with other commodities from two or more families, and a commodity can only belong to one family. Therefore, at most one family of

*e-mail: Z.Lin@leeds.ac.uk
**e-mail: R.S.Kwan@leeds.ac.uk

DOI: 10.1051/itmconf/20171400003

commodities can be used at a node/arc. In this paper, we focus on the case that the compatibility relations are imposed on nodes on a directed acyclic graph (DAG) as exhibited in the Train Unit Scheduling Problem (TUSP) [1, 2]. Some conclusions are applicable to a general directed graph by analogy.

Given a train operator's timetable and a fleet of train units, the TUSP aims to cover each timetabled train service by a train unit or units. The Train Unit Assignment Problem (TUAP) [9] is also relevant. In [9] and [2], integer multicommodity flow models are used where nodes represent trips to be covered by train units as commodities of different types. A notable feature of the TUSP is that more than one train unit can be coupled to serve the same trip, which gives the requirement on unit type compatibility relations, as often in practice, not any arbitrary train unit types can be coupled.

In [1, 2], in forming feasible train unit schedules, the above unit type compatibility requirement is satisfied by either creating additional variables or by customized branching. We generalize this class of problems as (I)MCFP-CCR. We give complexity analysis and propose recently developed solution methods. Computational experiments will be reported. Conclusion and future research are discussed finally.

2 (I)MCFP-CCR Models

Let $G = (N, A)$ be a DAG with commodities $k \in K, |K| > 1$ partitioned by families $f \in F$ and commodity-graphs $G^k = (N^k, A^k)$ as sub-graphs of G. Let P^k (P^k_j) be the set of commodity k paths (passing through node j). The path formulation of (I)MCFP-CCR with path variables x_p can be "formulated" by the following "(I)LP":

$$(PF) \quad \min \quad \sum_{k\in K} \sum_{p\in P^k} c_p x_p \tag{1}$$

$$\sum_{p\in P^k} x_p = b^k, \quad \forall k \in K \tag{2}$$

$$\sum_{k\in K_j} \sum_{p\in P^k_j} q^k x_p \geq r_j, \quad \forall j \in N \tag{3}$$

$$\sum_{k\in K_j} \sum_{p\in P^k_j} v^k x_p \leq u_j, \quad \forall j \in N \tag{4}$$

$$\text{used commodities at } j \text{ are compatible}, \quad \forall j \in N \tag{5}$$

$$x_p \in \mathbb{R}_+(\mathbb{Z}_+), \quad \forall p \in P^k, \forall k \in K \tag{6}$$

In (1), c_p is the cost for path p in P^k. (2) sets the total demand for each commodity. (3) gives the minimum supply requirement $r_j \geq 0$ where K_j is the set of allowed commodities at j and q^k is the "task contribution" by a unit flow of k. (4) gives the capacity upper bound $u_j \geq 0$, where v^k is the "capacity consumption" by a unit flow of k. "Constraints" (5) are only descriptive. They require used commodities at each node to be compatible. Finally (6) gives the variable domains for the continuous and integer variants respectively.

3 Computational Complexity

Proposition 1. *The IMCFP-CCR is strongly $\mathcal{NP}$-hard.*

Proof. Any TUAP [9] instance is a special case of IMCFP-CCR with only one family of commodities. TUAP is strongly $\mathcal{NP}$-hard [9]. Therefore, IMCFP-CCR is strongly $\mathcal{NP}$-hard. □

Next, it is shown that the decision problem of the MCFP-CCR, denoted as MCFP-CCR(D), is a reduction from an $\mathcal{NP}$-complete problem 3SAT [10]. Suppose there are n boolean variables $X = \{x_1, \ldots, x_n\}$ and m clauses $C_1, \ldots, C_m$, 3SAT asks whether there exists an assignment to X such that $C_1 \wedge \cdots \wedge C_m = T$, where each clause C_j is a conjunction of three logic variables such that each is either a variable in X or the negation of a variable in X. For example, $C_1 = x_1 \vee x_2 \vee \neg x_3$ is a possible clause.

Proposition 2. *The MCFP-CCR is $\mathcal{NP}$-hard.*

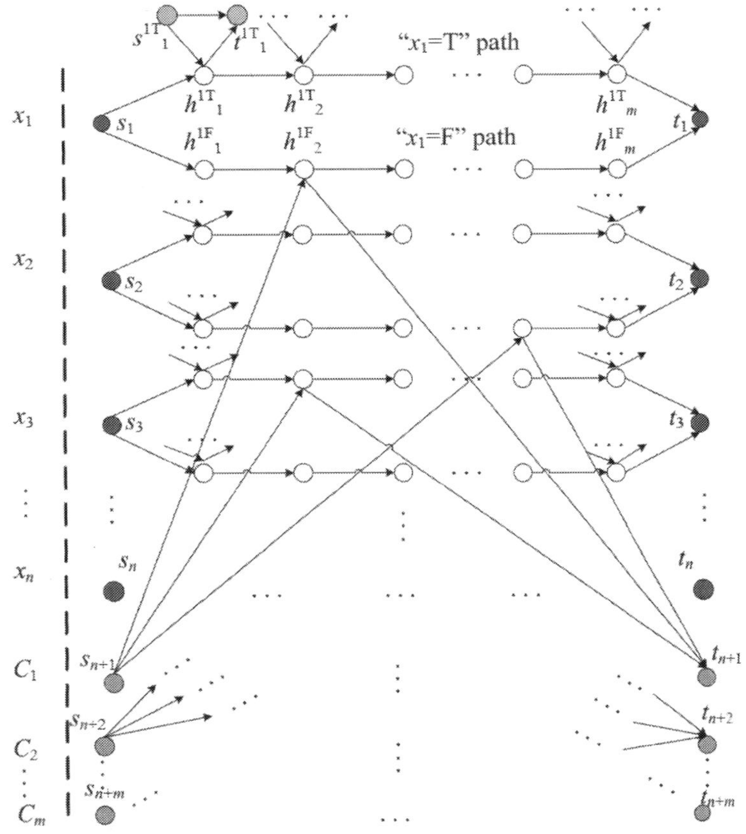

Figure 1: An illustration of an MCFP-CCR(D) instance in the proof, and highlights on dummy sources/sinks wrt node h_1^{1T} and clause $C_1 = x_1 \vee x_2 \vee \neg x_3$.

Proof. (Sketch) We show that any 3SAT problem can be polynomially reduced to a special MCFP-CCR(D) where each family has only one commodity, as illustrated in Fig. 1. For every boolean variable $x_i \in X$, we create a family source s_i and a family sink t_i, as well as a "true" path and a "false" path each containing m nodes, linked with s_i and t_i. For each of the $2m$ nodes, we add a dummy family source/sink pair and arcs as shown in Fig. 1. For very clause $C_j, j = 1, \ldots, m$, we create a family source s_{n+j} and a family sink t_{n+j}. Three arcs are created from s_{n+j} and three arcs are linked to t_{n+j} in the following way. If x_i is in C_j as itself, we add an arc from s_{n+j} to a node in the "false" path between s_i and t_i and another arc from this node to t_{n+j}; if x_i is in C_j as its negation, we add an arc from s_{n+j} to a node in the "true" path between s_i and t_i and another arc from this node to t_{n+j}. If more than one clauses use the same variable, their arcs should be connected to distinct nodes in the relevant paths. All family sources/sinks have a demand of 1, and for each internal node its r and u values are also set as 1. It can be verified that any 3SAT problem can be polynomially transformed to an MCFP-CCR(D) as described above. □

4 Solution Approaches: Node-family Branching

In [2] a method is proposed for the TUSP by adding extra binary variables associated for train-family associations. One disadvantage of the above node-family variables is the difficulty in incorporating them into a local convex hull scheme [11]. To resolve this, in [1, 11], an alternative way for ensuring family compatibility by branching is used. We generalize this idea as *node-family branching* and present ongoing research. It is suitable for both IMCFP-CCR and MCFP-CCR.

Let N' be the set of nodes that can be potentially covered by multiple families and let F_j be the set of families allowed at j. Denote the formulation by removing (5) from (PF) as (PF_0). Node-family branching removes incompatible commodities at nodes by branch-and-bound (BB). Suppose linear programming (LP) relaxation is used for getting lower bounds. The solution of the LP relaxation of (PF_0) at the current BB tree node will be checked and either (i) a network node covered by more than one family is identified or (ii) no such a network node can be found and node-family branching is stopped. Suppose a "multi-family" network node $j^* \in N'$ is detected with families $f_1, \ldots, f_r$ covering it, and a family $f^* \in \{f_1, \ldots, f_r\}$ is selected. Then two branches will be formed in the following way: (assuming the left has a higher priority)

- Left branch: only f^* is allowed to cover j^*.
- Right branch: f^* is forbidden to cover j^*.

The following discusses two technical issues regarding node-family branching. The first is on branching variable selection, that is, which (j^*, f^*) pair will be selected. The second is only relevant to the IMCFP-CCR, which investigates how to integrate node-family branching with integer branching.

4.1 Variable selection on (j^*, f^*)

Branch on the Most/Least Diverse Node.

This corresponds to branching on the "most/least fractional" (decimal part closest/farthest to 0.5) schemes used in fractional-to-integer branching. Given an LP relaxation of the tree node n_{BB} to be branched, we define the flow proportion of family f over a network node j as

$$R_{f,j} := \frac{\sum_{k \in f} \sum_{p \in P_j^k} x_p}{\sum_{k \in K_j} \sum_{p \in P_j^k} x_p}. \tag{7}$$

Then the *family diversity* of node j can be defined following Shannon's diversity index [12] that is commonly used as a quantitative measure in reflecting how many and how diverse of different types in a set of collection of them. The family diversity over node j is defined as

$$H_j = - \sum_{f \in F_j} R_{f,j} \ln R_{f,j}. \tag{8}$$

DOI: 10.1051/itmconf/20171400003

When all used families have the same proportion $\frac{1}{|F_j|}$, the family diversity takes its maximum value $H_j^{\max} = \ln |F_j|$. The more "unequal" the abundances of families, the smaller the value of H_j. When there is only one family, H_j takes its minimum value $H_j^{\min} = 0$. Then j^* can be selected either by $j^* = \arg\max_{H_j>0}(H_j)$, or its opposite $j^* = \arg\min_{H_j>0}(H_j)$. In either case, when j^* is chosen, we choose f^* by $f^* = \arg\max(R_{f,j^*})$.

Branching on Fixed Order.

An ordered list of all nodes in N' is set before the BB process as: $L(N') : \langle j_1, j_2, \ldots, j_{|N'|}\rangle$. The selection can be simply accomplished by finding the first multi-family covered node in the list. It avoids looping over the entire N' at the price of poor flexibility. This naive strategy can be especially useful for DAG, provided the list is constructed according to the "topological sorting" [3] of the nodes. The "single-direction" nature of a DAG suggests that a branching decision made on a node will often have significant impact on its successors.

Branching on Pseudocosts.

Pseudocost branching [13–16] tries to estimate the objective changes caused by branching on potential candidates and chooses the one yielding a largest change, based on empirical evidence that the pseudocost of each variable is relatively "constant". Although without explicit variables, the idea can be applied to node-family branching. Suppose at tree node n, the actual down and up pseudocosts due to branching on (j, f) can be calculated by

$$\psi_{j,f}^{n-} = \frac{z_{j,f}^{n-} - z^n}{R_{j,f}^n}, \quad \psi_{j,f}^{n+} = \frac{z_{j,f}^{n+} - z^n}{1 - R_{j,f}^n}, \tag{9}$$

where $R_{j,f}^n$ is the flow proportion at n as defined in (7), z^n is the objective value of n, and $z_{j,f}^{n-}$ and $z_{j,f}^{n+}$ are the objective values of the right and left branches if (j, f) is chosen. By collecting sufficient information of (j, f) over tree nodes $n_1, ..., n_M$, the "artificial" pseudocosts of (j, f), denoted by $\Psi_{j,f}^-, \Psi_{j,f}^+$, can be estimated from $\psi_{j,f}^{n_1-}, ..., \psi_{j,f}^{n_M-}$ and $\psi_{j,f}^{n_1+}, ..., \psi_{j,f}^{n_M+}$ respectively. Then the objective change at a subsequent tree node n can be estimated by

$$\Delta_{j,f}^{n-} = \Psi_{j,f}^- R_{j,f}^n, \quad \Delta_{j,f}^{n+} = \Psi_{j,f}^+(1 - R_{j,f}^n). \tag{10}$$

Finally, j^*, f^* can be determined at n by some rules. For example, applying the rules in [15], we have $(j^*, f^*) = \arg\max\{\alpha_1 \min(\Delta_{j,f}^{n-}, \Delta_{j,f}^{n+}) + \alpha_2 \max(\Delta_{j,f}^{n-}, \Delta_{j,f}^{n+})\}$, $\alpha_1 + \alpha_2 = 1, 0 < \alpha_1, \alpha_2 < 1$, with α_1, α_2 set from experiences.

4.2 IMCFP-CCR: Integration with Integer Branching

One way to solve the IMCFP-CCR is to integrate node-family branching with integer branching on the same BB tree. Since there are two branching systems, a priority arrangement has to be set if both of them can be used. We use "nf>int" to denote the case when node-family branching (nf) has priority over integer branching (int). This priority can be updated either in a *static* or *dynamic* way. In either case, an initial priority is set. For the static strategy, the priority remains the same throughout the BB process. For the dynamic strategy, the priority is updated such that if no branching candidate can be found according to the current branching system, the secondary system will take over from the current.

DOI: 10.1051/itmconf/20171400003

5 Computational Experiments

Real-world datasets based on a UK rail network were used. Two coupling incompatible train unit types, c360 and c321, differing in seat capacities and coupling upper bounds, were allowed to serve all trips. The objective is to minimize the number of used units. Constraints (3) and (4) were replaced by train convex hull constraints [1, 11]. A customized branch-and-price solver was used based the simplex solver of Xpress-MP 7.7. The maximum number of BB tree nodes was set to 5000 and the maximum time was set to 24 hours.

Table 1. Experiments on MCFP-CCR

instance	(j^*, f^*) selection	opt feas obj val	unit # (360/321)	feas sol #	BB#	time (second)	solution quality
	"one-family"	*23*	*8/15*	–	–	*2*	*optimal*
small network	fixed: toplg-sort	23	8/15	3	448	126	optimal
(137 nodes,	fixed: random	fail	–	0	5000	6637	–
2690 arcs)	**most diverse**	**23**	**8/15**	**2**	**101**	**24**	**optimal**
	least diverse	24	13/11	2	5000	11426	sub-opt
	"one-family"	*133*	*33/100*	–	–	*79*	*optimal*
large network	fixed: toplg-sort	134	34/100	5	445	86400	sub-opt
(512 nodes,	fixed: random	fail	–	0	1012	86400	–
23015 arcs)	most diverse	137	37/100	2	910	86400	sub-opt
	least diverse	**133**	**33/100**	**2**	**181**	**592**	**optimal**

Table 1 gives the results for MCFP-CCR on a small and a large network respectively. Four train-family selection strategies and an instance where the two families were set to be compatible ("one-family") were compared. There is evidence that topological sorting based on trains' departure times is advantageous compared with branching on a random order of the trains. Most/least diverse branching outperformed the rest. Table 2 gives the results on experiments for IMCFP-CCR on the small network. Different branching strategies were compared. The initial priority setting can be crucial as all the runs with "int>tf" except one failed. The differences between static and dynamic strategies were not very significant. The "most diverse" train-family selection strategy outperformed the rest. Experiments regarding large scaled network IMCFP-CCR and the selection strategy based on pseudocosts will be conducted in the future.

Acknowledgments. This research is supported by an EPSRC project EP/M007243/1. We would like to also thank FirstGroup for their collaboration.
Data Statement FirstGroup has provided relevant data for the research, part of which is commercially sensitive. Where possible, the data that can be made publicly available is deposited in `http://archive.researchdata.leeds.ac.uk/`.

References

[1] Z. Lin, R.S.K. Kwan, Electron. Notes Discrete Math **41**, 165 (2013)

[2] Z. Lin, R.S.K. Kwan, Public Transport **6**, 35 (2014)

[3] R.K. Ahuja, T.L. Magnanti, J.B. Orlin, *Network flows: theory, algorithms and applications* (Prentice Hall, Englewoods Cliffs, USA, 1993)

[4] L.R. Ford, D.R. Fulkerson, Manag. Sci. **5**, 97 (1958)

DOI: 10.1051/itmconf/20171400003

Table 2. Experiments on IMCFP-CCR: small network (137 nodes, 2690 arcs)

initial priority	(j^*, f^*) selection	update strategy	opt feas obj val	unit # (360/321)	feas sol #	BB#	time (second)	rule % (tf/int)	solution quality
"one-family"	–	–	*23*	*9/14*	–	*169*	*58*	–	*optimal*
tf>int	fixed: toplg-sort	dynamic	23	8/15	3	1441	858	94/6	optimal
tf>int	fixed: toplg-sort	static	23	8/15	3	1449	873	94/6	optimal
tf>int	fixed: random	dynamic	26	11/15	1	5000	8000	100/0	sub-opt
tf>int	fixed: random	static	24	9/15	3	5000	24268	99/1	sub-opt
tf>int	most diverse	dynamic	23	8/15	2	585	426	75/25	optimal
tf>int	**most diverse**	**static**	**23**	**8/15**	**2**	**197**	**127**	**50/50**	**optimal**
tf>int	least diverse	dynamic	24	13/11	2	5000	20962	85/15	sub-opt
tf>int	least diverse	static	24	13/11	2	5000	11901	98/2	sub-opt
int>tf	fixed: toplg-sort	dynamic	fail	–	0	5000	9217	4/96	–
int>tf	fixed: toplg-sort	static	fail	–	0	5000	5991	6/94	–
int>tf	fixed: random	dynamic	fail	–	0	5000	9458	97/3	–
int>tf	fixed: random	static	fail	–	0	5000	9005	8/92	–
int>tf	most diverse	dynamic	fail	–	0	5000	14823	97/3	–
int>tf	most diverse	static	fail	–	0	5000	14406	9/91	–
int>tf	least diverse	dynamic	25	10/15	1	5000	4237	96/4	sub-opt
int>tf	least diverse	static	fail	–	0	5000	10612	4/96	–

[5] T.C. Hu, Oper. Res. **11**, 344 (1963)
[6] J.L. Kennington, Oper. Res. **26**, 209 (1978)
[7] R.M. Karp, Networks **5**, 45 (1975)
[8] S. Even, A. Itai, A. Shamir, SIAM J. Comput. **5**, 691 (1976)
[9] V. Cacchiani, A. Caprara, P. Toth, Math. Prog. B **124**, 207 (2010)
[10] M.R. Garey, D.S. Johnson, *Computers and Intractability: A Guide to the Theory of NP-Completeness* (W. H. Freeman & Co., New York, NY, USA, 1979), ISBN 0716710447
[11] Z. Lin, R.S.K. Kwan, Comput. Optim. Appl. pp. 1–39 (2016)
[12] C.E. Shannon, Bell System Technical Journal **27**, 379 (1948)
[13] M. Benichou, J.M. Gauthier, P. Girodet, G. Hentges, G. Ribiere, O. Vincent, Math. Prog. **1**, 76 (1971)
[14] J.M. Gauthier, G. Ribière, Math. Prog. **12**, 26 (1977)
[15] J.T. Linderoth, M.W.P. Savelsbergh, INFORMS J. on Comput **11**, 173 (1999)
[16] T. Achterberg, T. Koch, A. Martin, Oper. Res. Letters **33**, 42 (2005)

Named Entity Recognition: Resource Constrained Maximum Path

Luigi Di Puglia Pugliese[2,*], *Elisabetta* Fersini[1,**], *Francesca* Guerriero[2,***], and *Enza* Messina[1,****]

[1] *University of Milano-Bicocca*
[2] *University of Calabria*

Abstract. Information Extraction (IE) is a process focused on automatic extraction of structured information from unstructured text sources. One open research field of IE relates to Named Entity Recognition (NER), aimed at identifying and associating atomic elements in a given text to a predefined category such as names of persons, organizations, locations and so on. This problem can be formalized as the assignment of a finite sequence of semantic labels to a set of interdependent variables associated with text fragments, and can modelled through a stochastic process involving both hidden variables (semantic labels) and observed variables (textual cues). In this work we investigate one of the most promising model for NER based on Conditional Random Fields (CRFs). CRFs are enhanced in a two stages approach to include in the decision process logic rules that can be either extracted from data or defined by domain experts. The problem is defined as a Resource Constrained Maximum Path Problem (RCMPP) associating a resource with each logic rule. Proper resource Extension Functions (REFs) and upper bound on the resource consumptions are defined in order to model the logic rules as knapsack-like constraints. A well-tailored dynamic programming procedure is defined to address the RCMPP.

1 Introduction

Information Extraction (IE) is a task of Natural Language Processing aimed at inferring a structured representation of contents from unstructured textual sources. In this field, Named Entity Recognition (NER) has gained the attention of researches for identifying and associating atomic elements in a given text to a predefined category (such as names of persons, organizations and locations). Considering a text as sequence of tokens $x = x_1, \ldots, x_N$, the goal is

*e-mail: luigi.dipugliapugliese@unical.it
**e-mail: fersini@disco.unimib.it
***e-mail: francesca.guerriero@unical.it
****e-mail: messina@disco.unimib.it

to classify each token x_i as one of the entity labels $y_j \in Y$ for originating a tag sequence $y = y_1, \ldots, y_N$.

Nowadays, the state-of-the-art model for tackling the NER task is represented by linear chain Conditional Random Fields (CRFs) [3], which is a discriminative undirected graphical model able to encode known relationships among tokens (observations) and labels (hidden states). In order to efficiently enhance the description power of CRFs, two main research directions have been investigated to enlarge the information set exploited during training and inference: (1) relaxing the Markov assumption [6] to include long distance dependencies and (2) introducing additional domain knowledge in terms of logical constraints [2, 5]. Considering that the relaxation of the Markov assumption implies an increasing computational complexity, in this paper we focused on the second research direction by formulating the inference task as an Integer Linear Programming problem. In particular, the standard CRFs inference process is enhanced by two main contribution: (1) introducing "extra knowledge" related to semantic constraints about the token labels, and (2) modelling the label assignment problem as a Resource Constrained Maximum Path Problem (RCMPP).

2 Conditional Random Fields

2.1 Background

A CRF [3] is an undirected graphical model that defines a single joint distribution $P(y|x)$ of the predicted labels (hidden states) $y = y_1, ..., y_N$ given the corresponding tokens (observations) $x = x_1, ..., x_N$. Linear Chain CRFs, in which a first-order Markov assumption is made on the hidden variables, define the following conditional distribution:

$$p(\vec{y}|\vec{x}) = \frac{1}{Z(\vec{x})} \exp\Big(\sum_{t=1}^{N} \sum_{k}^{K} \omega_k f_k(y_t, y_{t-1} x, t)\Big) \tag{1}$$

where $f_k(y_t, y_{t-1} x, t)$ is an arbitrary real-valued feature function over its arguments and ω_k is a learned weight that tunes the importance of each feature function. In particular, when for a token x_t a given feature function f_k is active, the corresponding weight ω_k indicates how to take into account f_k: (1) if $\omega_k > 0$ it increases the probability of the tag sequence y; (2) if $\omega_k < 0$ it decreases the probability of the tag sequence y; (3) if $\omega_k = 0$ has no effect whatsoever. Once the parameters ω_k have been estimated, usually by maximizing the likelihood of the training data [4], the inference phase can be addressed.

2.2 Inference: finding the most probable state sequence

The inference problem in CRF corresponds to find the most likely sequence of hidden state y^*, given the set of observation $x = x_1, ..., x_n$. This problem can be solved by determining y^* such that:

$$y^* = \arg\max_{y} p(\vec{y}|\vec{x}) \tag{2}$$

DOI: 10.1051/itmconf/20171400004

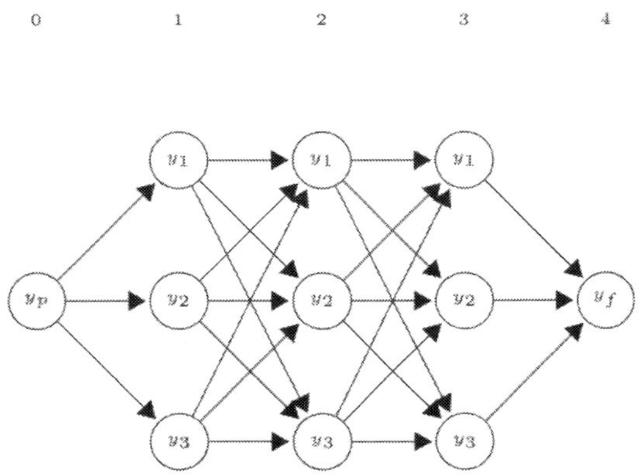

Figure 1. Layered acyclic directed graph.

Given a number m of possible states and n possible input tokens, a layered acyclic directed graph D can be constructed for addressing the inference problem. The graph D is composed of $n+1$ layers. Layer 0 corresponds to the entry layer, $n+1$ is the ending layer, the other n layers represent the elements of the sequence. Arcs from each state y_i, $i = 1, \ldots, m$ belonging to each layer t exists for each state y_i, $i = 1, \ldots, m$ belonging to layer $t+1$, $t = 0, \ldots, n$. We denote as $\mathcal{N}$ the set of nodes containing $2 + n \times m$ elements, that is, the states y_i, $i = 1, \ldots, m$ associated with all layers $t = 1, \ldots, n$ and two additional states named start state y_p and final state y_f. The set $\mathcal{A}$ contains the arcs (t, y_i, y_i'), $t = 0, \ldots, n$, $i = 1, \ldots, m$. Arc (t, y_i, y_i') denotes the link of state y_i associated with layer t with the state y_i' associated with layer $t+1$.

A value $\alpha^t_{y_i, y_i'}$ is associated with each arc $(t, y_i, y_i') \in \mathcal{A}$. The parameter $\alpha^t_{y_i, y_i'}$ is proportional to the probability of passing from state y_i at layer t to state y_i' at layer $t+1$. Figure 1 shows the graph D with $n = m = 3$.

Given a layered acyclic directed graph denoting both observed tokens and labels to be assigned, the objective is to find the heaviest path in the graph D starting from y_p and ending at the state y_f. Given the variables $e^t_{y_i, y_i'}$ assuming value equal to 1 if arc (t, y_i, y_i') is included to the optimal path, 0 otherwise, the problem can be formulated as follows.

$$\max \sum_{(t, y_i, y_i') \in \mathcal{A}} \alpha^t_{y_i, y_i'} e^t_{y_i, y_i'} \tag{3}$$

$s.t.$

$$\sum_{(t-1, y_i, y_i') \in \mathcal{A}} e^{t-1}_{y_i, y_i'} - \sum_{(t, y_i', y_i) \in \mathcal{A}} e^t_{y_i', y_i} = 0, \ \forall y_i' \in N \setminus \{y_p, y_f\},$$

$$1 \leq t \leq n \tag{4}$$

$$\sum_{(0, y_p, y_i)} e^0_{y_p, y_i} = 1 \tag{5}$$

$$\sum_{(n, y_i, y_f)} e^n_{y_i, y_f} = 1 \tag{6}$$

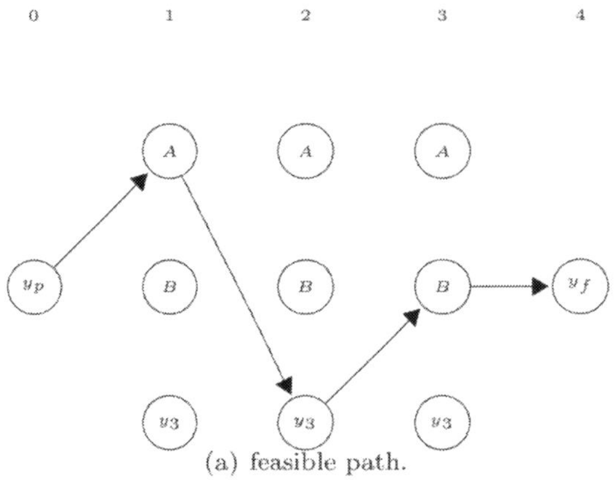

(a) feasible path.

Figure 2. Feasible paths for constraints (8).

The following constraints, opportunely instantiated according to the domain knowledge, could be introduced in the model:

- Adjacency: if the token at time $t-1$ is labelled as A, then the token at time t must be labelled as B

$$\sum_{y_i \in \mathcal{N}} e^{t-1}_{y_i,A} - e^{t}_{A,B} \leq 0, \quad t = 1, \ldots, n-1. \tag{7}$$

- Precedence: if the token at time $t+z$ is labelled as B, then a token at time t must be labelled as A

$$\sum_{y_i \in \mathcal{N}} e^{t-1}_{y_i,A} - \sum_{z=1}^{n-t-1} \sum_{y_i \in \mathcal{N}} e^{t+z}_{B,y_i} \geq 0, \quad t = 1, \ldots, n-1. \tag{8}$$

- Begin-end position: if the sequence of tokens starts with label A, then the sequence must end with label B

$$e^{0}_{y_p,A} - e^{n}_{B,y_f} \leq 0. \tag{9}$$

To guarantee constraints (7) it is sufficient to modify the graph (D) by removing all the edges (t, y_i, B). Examples of feasible paths satisfying constraints (8) are depicted in figure 2. Figure 3 shows an infeasible path for the same constraints.

3 Resource constrained model

The problem can be modelled as a Resource Constrained Maximum Path Problem (*RCMPP*). It is possible to define proper Resource Extension Function (*REF*) in order to introduce knapsack-like constraints for each typology of logic rule, that is, precedence and begin-end conditions. The reader is referred to [1] for more details on resource constrained path problems.

 DOI: 10.1051/itmconf/20171400004

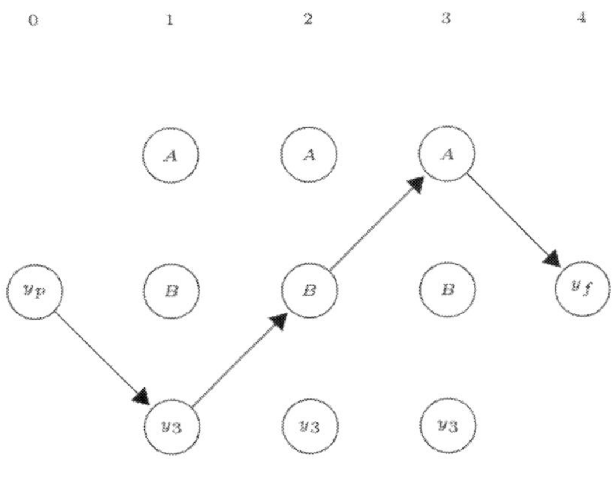

Figure 3. Unfeasible path for constraints (8).

3.1 Precedence Constraint

Let consider P_B be the set of predecessor states of state B. The path from y_p to B has to contain all states $y_{\bar{i}} \in P_B$. We assume a resource consumption $r^P_{i,\bar{i}}$ associated with each state $y_{\bar{i}}$, for each y_i of the network, defined in what follows:

$$r^P_{i,\bar{i}} = \begin{cases} 1, & if \;\; y_i \equiv B; \\ -1, & if \;\; i = \bar{i}; \\ 0, & if \;\; i \neq \bar{i}, y_i \not\equiv B. \end{cases} \tag{10}$$

The REF associated with the precedence constraints is defined in Eq. (11).

$$\begin{cases} w^P_{y_p,\bar{i}} = 0, \forall y_{\bar{i}} \in P_B; \\ w^P_{j,\bar{i}} = w^P_{i,\bar{i}} + r^P_{j,\bar{i}}, \forall (t, y_i, y_j) \in \mathcal{A}, y_{\bar{i}} \in P_B \cup \{B\}. \end{cases} \tag{11}$$

The resource limit W_{P_B} is set equal to 0. The set of knapsack-like constraints that define the precedence constraints assume the following form:

$$w^P_{j,\bar{i}} \leq W_{P_B}, \forall j \in \mathcal{N}, \forall y_{\bar{i}} \in P_B. \tag{12}$$

Figure 4 shows the resource constrained instance when precedence constraints are considered.

Considering feasible paths in figure 2, the resource consumption is either 0 or -1 for each label of both paths. The path in Figure 3 has resource consumption equal to 1 at state B, thus it is infeasible for Eq. (12).

It is worth observing that the resource at states can be viewed as resource on arcs. Indeed, the resource consumption of an arc is the resource consumption of the head node.

3.2 Begin-end Constraint

Here we define resource constraint for the begin-and condition. We assume a resource consumption $r^{be}_{(t,y_i,y_j)}$ associated with each arc $(t, y_i, y_j) \in \mathcal{A}$.

 DOI: 10.1051/itmconf/20171400004

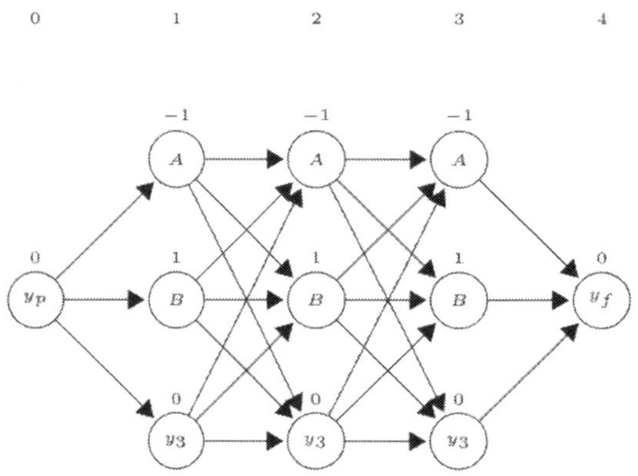

Figure 4. Resource constrained graph with the precedence constraint $P_B = \{A\}$.

$$r^{be}_{(t,y_i,y_j)} = \begin{cases} 1, & if\ \ t = 0, y_i = y_p, y_j = A; \\ -1, & if\ \ t = n, y_i = B, y_j = y_f; \\ 0, & for\ all\ other\ cases. \end{cases} \tag{13}$$

The REF associated with begin-and constraint is defined in equation (14).

$$\begin{cases} w^{be}_{y_p} = 0; \\ w^{be}_j = w^{be}_i + r^{be}_{(t,y_i,y_j)}, \forall (t, y_i, y_j) \in \mathcal{A}. \end{cases} \tag{14}$$

Assuming a resource limit $W_{be} = 0$, the resource constraint modelling begin-end condition is expressed by equation (15).

$$\sum_{(t,y_i,y'_i)\in\mathcal{A}} e^t_{y_i,y'_i} r^{be}_{(t,y_i,y'_i)} \leq W_{be}. \tag{15}$$

Figure 5 shows the resource graph associated with the begin-end condition.

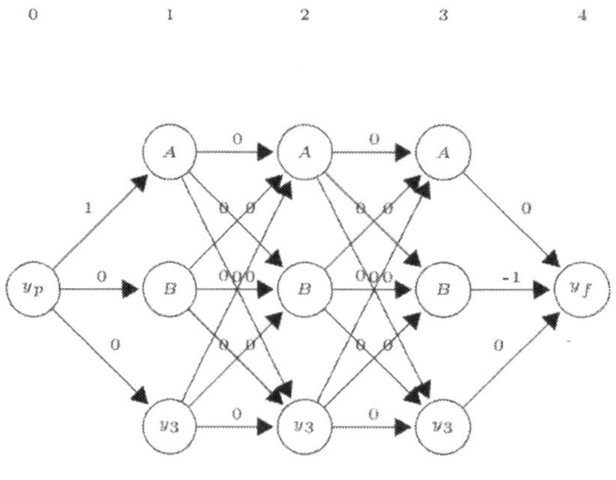

Figure 5. Resource constrained graph with begin-end condition.

4 Sketch of dynamic programming

Let l_i^t be a label associated with state y_i at layer t. The label l_i^t maintains information related to the objective function (α_i^t) and to the resource consumption $(w_{i,\bar{i}}^P, w_i^{be})$. A path π_i^t is associated with each label $l_i^t(\alpha_i^t, w_{i,\bar{i}}^P, w_i^{be})$. Since there may exist several paths to state y_i, the index h is used to indicate the id of a given path. Thus, label $l_i^t(h)$ is associated to the $h-th$ path $\pi_i^t(h)$. Starting from the initial label $l_{y_p}^0(0,0,0)$, the dynamic programming explores the state-space in order to reach the final labels $l_{y_f}^n(h)(\cdot)$. Among all labels $l_{y_f}^n(h)(\cdot)$, that with maximum value of α is associated with optimal path. The state-space is reduced by considering only feasible and non-dominated labels.

Definition 4.1 (Feasibility). A label $l_i^t(h)$ is feasible if and only if $w_{i,\bar{i}}^P(h) \leq W_{P_B}, \forall y_{\bar{i}} \in P_B$ and $w_i^{be}(h) \leq W_{be}$, with $i \equiv y_f$.

Definition 4.2 (Dominance). Given two labels $l_i^t(h)$ and $l_i^t(h')$, the first dominates the second if the following conditions hold

$$\alpha_i^t(h) \geq \alpha_i^t(h'); \tag{16}$$

$$w_{i,\bar{i}}^P(h) \leq w_{i,\bar{i}}^P(h'); \tag{17}$$

$$w_i^{be}(h) \leq w_i^{be}(h'); \tag{18}$$

and at least one inequality is strictly satisfied.

Let L be the list of labels with the potential to generate an optimal solution and ND_i be the set of non-dominated labels associated with y_i. The labelling approach to solve to optimality the problem is depicted in Algorithm 1.

Algorithm 1 . $\mathcal{DP}$ scheme

```
Step 0 (Initialization Phase)
Set: L = {l^0_{y_p}(0)}.

Step 1 (Label Selection)
Select and delete from L a label l^t_i(h).

Step 2 (Labels Generation)
for all y_j : (t, y_i, y_j) ∈ A do
   Compute α^{t+1}_j, w^P_{j,ī}, ∀y_ī ∈ P_B, and w^{be}_j.
   if Label l^{t+1}_j(α^{t+1}_j, w^P_{j,ī}, w^{be}_j) is feasible then
      if Label l^{t+1}_j is not dominated by any labels in ND_j then
         Add label l^{t+1}_j to ND_j and L.
         Remove from ND_j all labels dominated by l^{t+1}_j.
      end if
   end if
end for

Step 3 (Termination check)
if L = ∅ then
   STOP
else
   Go to Step 1.
end if
```

5 Conclusion

In this paper, the problem of Named Entity Recognition is addressed by investigating the inference task on Conditional Random Fields. In particular, a mathematical programming formulation based on a Resource Constrained Maximum Path is presented to include some background knowledge during the labelling phase of a text source. Three types of background knowledge constraints have been presented, together with a dynamic programming approach for determining the optimal sequence of labels. Concerning the future work, additional long distance dependencies are planned to be automatically discovered from the data (as hidden patterns) and enclosed into the mathematical problem formulation.

References

[1] L. Di Puglia Pugliese and F. Guerriero. A survey of resource constrained shortest path problems: Exact solution approaches. *Networks*, 62(3):183–200, 2013.

[2] E. Fersini, E. Messina, G. Felici, and D. Roth. Soft-constrained inference for named entity recognition. *Information Processing & Management*, 50(5):807–819, 2014.

[3] J. D. Lafferty, A. McCallum, and F. C. N. Pereira. Conditional random fields: Probabilistic models for segmenting and labeling sequence data. In *Proc. of the 18th Int. Conf on Machine Learning*, 2001.

[4] R. Malouf. A comparison of algorithms for maximum entropy parameter estimation. In *Proc. of the 6th Conf. on Natural language learning*, 2002.

[5] D. Roth and W.-T. Yih. Ilp inference for conditional random fields. In *Proc. of the 22nd Int. Conf. on ML*, 2005.

[6] S. Sarawagi and W. W. Cohen. Semi-markov conditional random fields for information extraction. In *Proc. of the 18th Conf. on Neural Information Processing Systems*, 2004.

 DOI: 10.1051/itmconf/20171400005

MIP models for the irregular strip packing problem: new symmetry breaking constraints

Marcos Okamura Rodrigues[1,*], *Luiz Henrique* Cherri[1,**], and *Leandro Resende* Mundim[1,***]

[1] *Instituto de Ciências Matemáticas e de Computação, Universidade de São Paulo, Avenida Trabalhador São-carlense, 400 - Centro, São Carlos, Brazil*

Abstract. The irregular strip packing problem consists in minimizing the length used to cut a set of pieces from a board with fixed width. Recently, a mixed integer programming model was proposed for the problem, but it may allow a large number of symmetric solutions. In this paper, new symmetry breaking constraints are proposed to improve the model. Computational experiments were performed for instances with convex pieces. The results show the proposed formulation is better than the previous one for most instances, since it improves lower bounds and reduces run-time and number of nodes explored to prove optimality.

1 Introduction

The irregular strip packing problem (also known as irregular open dimensional problem) consists in cutting irregular pieces from a rectangular board with fixed width and infinite length. In the problem, pieces cannot overlap and each piece must be inside the board. The objective is to minimize the board length used to cut the pieces. The problem is both economical and environmental relevant, since reducing material consumption decreases production costs and minimizes wastes.

The irregular strip packing problem is NP-hard [9] and the methods proposed to solve it are mainly heuristics. The best solutions for large benchmark instances were found in [4]. He proposed a guided cuckoo search algorithm to solve the problem. In his method, the best fit between two pieces is found using a pairwise clustering and then the guided cuckoo search algorithm is applied to find complete solutions.

Some mixed integer programming (MIP) models were proposed to represent the problem. Fischetti and Luzzi [5] were the first to propose a MIP model to the problem. Alvarez-Valdes et al. [1] defined some structures used in [5] and developed a branch-and-bound algorithm to solve the model. Santoro and Lemos [11] proposed a MIP model to convex polygons with some limitations: the polygons can have up to eight edges and their opposite edges must be parallel. Cherri et al. [3] proposed two models that allow rotation of pieces. The models differ in the geometric structures used to avoid overlap among pieces. Toledo et al. [12] proposed the dotted-board model where the pieces were placed in a finite set of dots inside the board. The optimality of the model is subject to the

*e-mail: okamura@icmc.usp.br
**e-mail: lhcherri@icmc.usp.br
***e-mail: mundim@icmc.usp.br

discretization used to generate the dots. In order to combine the accuracy of continuous models with the advantages of the dotted-board model, Leao et al. [8] proposed a semi-continuous model where the pieces can be placed in continuous positions in x-axis and in discrete positions in y-axis.

Here, new symmetry breaking constraints for the MIP model described in [3] are proposed. Computational results show the performance of the model is improved in various aspects such as lower bounds, number of nodes explored and run-time.

2 MIP model

In this section, we define the irregular strip packing problem and review the No-Fit Polygon Covering Model (NFP-CM) proposed in [3].

Given a set of N pieces and a board with fixed width W, the irregular strip packing problem consists in minimizing the length z of the board used to cut all pieces. Each piece is defined by a set of vertices in clockwise orientation and one of the vertices is chosen to be its reference point. Let $w_i^{\min}$ ($w_i^{\max}$) be the vertical distance between the reference point of piece i and the vertex of the piece that is nearest (farthest) to the origin. Likewise, let $\ell_i^{\min}$ ($\ell_i^{\max}$) be the distance of the reference point of piece i and the leftmost (rightmost) vertex of the piece. Consider (x_i, y_i) the variables that define the translation of the reference point of piece i to place it in the board. Figure 1 illustrates the distance parameters of a piece in NFP-CM.

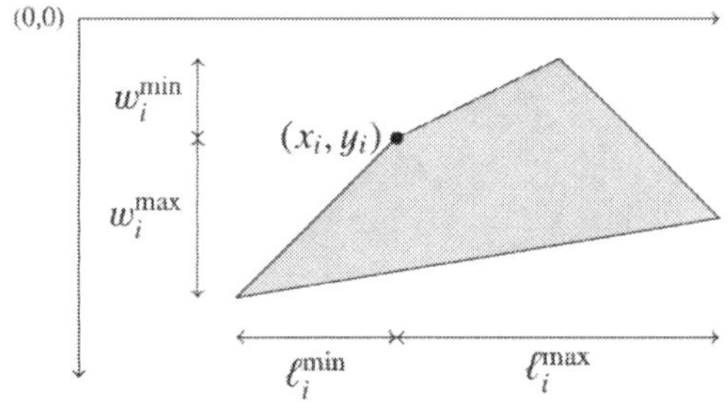

Figure 1: Representation of a piece and its distance parameters in NFP-CM.

The main difficulty to solve the problem is the non-overlap constraints. In NFP-CM the no-fit polygon is used to prevent overlap. For a review about non-overlap techniques and the geometry of the problem, see [2].

[3] defined the following binary variables to prevent overlap:

$$\xi_{ij}^k = \begin{cases} 1, & \text{if piece } j \text{ is on the right side of edge } k \text{ of } NFP_{ij}; \\ 0, & \text{otherwise.} \end{cases} \tag{1}$$

where NFP_{ij} is the no-fit polygon between pieces i and j. If piece i is on the right side of at least one edge $k = 1, ..., K_{ij}$ of NFP_{ij}, there is no overlap between pieces i and j.

NFP-CM is given by:

$$\min \quad z \tag{2}$$

 DOI: 10.1051/itmconf/20171400005

s.t.

$$\ell_i^{\min} \leq x_i \leq z - \ell_i^{\max}, \qquad i = 1, ..., N, \tag{3}$$

$$w_i^{\min} \leq y_i \leq W - w_i^{\max}, \qquad i = 1, ..., N, \tag{4}$$

$$a_{ij}^k(x_i - x_j) + b_{ij}^k(y_i - y_j) \leq c_{ij}^k + M_{ij}^k(1 - \xi_{ij}^k), \qquad 1 \leq i < j \leq N, k = 1, ..., K_{ij}, \tag{5}$$

$$\sum_{k=1}^{K_{ij}} \xi_{ij}^k = 1, \qquad 1 \leq i < j \leq N, \tag{6}$$

$$\xi_{ij}^k \in \{0, 1\}, \qquad 1 \leq i < j \leq N, k = 1, ..., K_{ij}. \tag{7}$$

Objective function (2) minimizes the length used to cut all pieces. Constraints (3) and (4) impose the pieces are placed inside the board. Constraints (5) prevent overlap among pieces. Constraints (6) ensure each piece is on the right side of an edge of the no-fit polygon while constraints (7) define the domain of binary variables.

3 Symmetry breaking constraints

NFP-CM (2)-(7) may have infinite symmetric solutions. Therefore, identical or equivalent solutions can be found several times by the enumeration scheme. In this section symmetric breaking constraints are proposed. The constraints can improve the linear relaxation of the model and reduce the number of identical solutions.

3.1 Partitioning the solution space

The main drawback of NFP-CM is non-overlap constraints (5). The definition of variables ξ_{ij}^k generates overlap between various regions, as shown in Figure 2(a). Dotted regions are common to two binary variables and hatched regions are common to three variables.

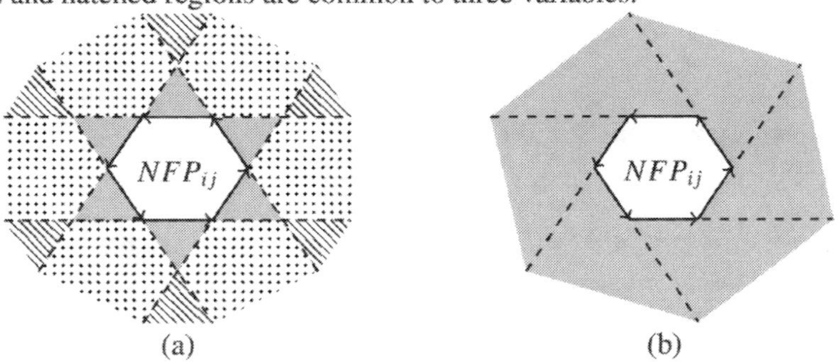

Figure 2: Non-overlap regions. (a) NFP-CM model. (b) Proposed partitioning.

Such redundancies on ξ_{ij}^k values can be avoided by adding constraints (8) to the model. The constraints ensure the reference point of piece j is on the left side of the line defined by edge k', where edge k' precedes edge k in the no-fit polygon. Therefore, the reference point of piece j can activate only one ξ_{ij}^k variable. Figure 2(b) illustrates the partitioning obtained from a no-fit polygon.

$$\overline{a}_{ij}^{k'}(x_i - x_j) + \overline{b}_{ij}^{k'}(y_i - y_j) \leq \overline{c}_{ij}^{k'} + \overline{M}_{ij}^{k'}(1 - \xi_{ij}^k), \quad 1 \leq i < j \leq N, k = 1, ..., K_{ij}. \tag{8}$$

3.2 Connecting pieces

In a feasible solution for the irregular strip packing problem, some pieces may not touch each other. However, one can translate the pieces to connect them, obtaining another solution with the same quality. Consider the following variables:

$$\phi_{ij} = \begin{cases} 1, & \text{if piece } i \text{ is connected to piece } j; \\ 0, & \text{otherwise.} \end{cases} \tag{9}$$

Constraints (10) combined with (5) ensure if ϕ_{ij} is equal to 1, the pieces i and j are connected.

$$a_{ij}^k(x_i - x_j) + b_{ij}^k(y_i - y_j) \geq c_{ij}^k - \underline{M}_{ij}^k(1 - \phi_{ij}), \quad 1 \leq i < j \leq N, k = 1, ..., K_{ij}. \tag{10}$$

To ensure each piece is connected with another piece, the following constraints are imposed:

$$\sum_{j=1}^{i-1} \phi_{ji} + \sum_{j=i+1}^{N} \phi_{ij} \geq 1, \quad i = 1, ..., N. \tag{11}$$

Note that constraints (11) do not ensure all pieces are connected as a single component. For example, it may have two groups of connected pieces with no interconnection. To enforce all pieces are connected, one can impose all cut sets of pieces are connected, as defined by constraints (12).

$$\sum_{(i,j) \in \delta(S) | i < j} \phi_{ij} \geq 1, \quad \forall S \subset N, S \neq \emptyset \tag{12}$$

where $\delta(S)$ is the cutset of S.

Although constraints (12) impose connectivity, they introduce an exponential number of constraints in the model and therefore hamper its resolution.

3.3 Connecting pieces with the board

In the optimal solution, at least one piece can touch the bottommost position of the board. In fact, if there is a solution that does not satisfy this statement, one can translate a piece down until it touches the board, obtaining a solution with the same quality.

Let ω_i be a binary variable, defined as follows:

$$\omega_i = \begin{cases} 1, & \text{if piece } i \text{ is in the bottommost position of the board;} \\ 0, & \text{otherwise.} \end{cases}$$

Constraints (13) and (14) ensure at least one piece is placed in the bottommost position in the board.

$$y_i \leq w_i^{min} + (W - w_i^{max} - w_i^{min})(1 - \omega_i), \quad i = 1, ..., N, \tag{13}$$

$$\sum_{i=1}^{N} \omega_i = 1. \tag{14}$$

3.4 Pieces with same shape

Symmetric solutions can be found if there are multiple pieces with same type (shape). For example, if there are two pieces with same type, all the solutions can be duplicated in an enumeration scheme. To avoid such problem, the pieces of same type can be ordered. Consider T_i the type of piece i, to eliminate the symmetry on the placement of pieces of same type, the following constraints are imposed.

$$y_i \leq y_j, \quad i = 1, ..., N, j = \min\{k = i+1, .., N | T_k = T_i\}. \tag{15}$$

Note that constraints (15), except in the case where the pieces are placed in the same y-position, eliminate symmetry of identical pieces.

Some values of ω_i can be inferred for pieces with same type using constraints (13)-(15). Specifically, for a piece with several copies, only the smallest index constraint (13) can be active. Constraints (16) ensure only the smallest index copy of each piece type can activate constraints (13).

$$\omega_j = 0, \quad i = 1, ..., N, j = \min\{k = i+1, .., N | T_k = T_i\}. \tag{16}$$

4 Results

In this section we present computational experiments to evaluate the performance of the proposed symmetry breaking constraints. The experiments were run on a computer with Intel® Core™ i7-2600 @ 3.40GHz CPU and 16 GB RAM under Ubuntu 14.04 LTS 64 bits operating system. The instances were solved by IBM ILOG CPLEX 12.6 solver using default parameters and a time limit of 3,600 seconds.

Table 1 presents the main characteristics of the evaluated instances: the name, the author(s) who proposed it, the total number of pieces, the number of piece types, the average number of vertices, and the upper bound (UB). The total number of pieces ranges between 3 and 17, with an average of 8.6 pieces, and an average number of 4.02 vertices per instance. Additionally, we present the upper bound (UB) used to estimate the big-M coefficients in constraints (5), for more details see [3]. UB was calculated using the bottom-left heuristic proposed in [6] with one thousand random input sequences.

Table 2 reports the computational experiments of NFP-CM proposed in [3] and the improved NFP-CM, i.e., NFP-CM with symmetry breaking constraints (8) and (13)-(16). Improved NFP-CM obtained better or equal lower bounds (LB) for all instances. Moreover, improved NFP-CM proved optimality for nineteen instances, four more instances than NFP-CM. In average, improved NFP-CM was 35.70% faster than NFP-CM to prove optimality. However, improved NFP-CM was slower for instance santoro4. Considering nodes explored by CPLEX, three instances were solved at the root node. For the remaining instances, improved NFP-CM explored fewer nodes for fourteen cases. The biggest reduction of nodes explored to prove optimality was obtained for instance santoro1, where NFP-CM explored 185.36 times the amount of nodes of improved NFP-CM. Furthermore, improved NFP-CM solved all instances santoro faster than the original paper [11]. We also highlight improved NFP-CM is the first MIP model to prove optimality for instances threep3, threep3w9 and fu in less than 1,200 seconds.

5 Conclusions and further work

In this paper we have proposed symmetry breaking constraints for the irregular strip packing problem. Although this new set of constraints was added to NFP-CM proposed in [3], the ideas presented here can be applied to other models.

Table 1: Features of benchmark instances.

Instance	Author(s)	Pieces	Piece types	Avg. vertices	UB
jones1	Jones [7]	4	4	3.25	5.00
jones2	Jones [7]	6	6	3.83	10.00
three	Alvarez-Valdes et al. [1]	3	3	3.67	6.67
threep2	Alvarez-Valdes et al. [1]	6	3	3.67	9.67
threep2w9	Alvarez-Valdes et al. [1]	6	3	3.67	8.67
threep3	Alvarez-Valdes et al. [1]	9	3	3.67	14.00
threep3w9	Alvarez-Valdes et al. [1]	9	3	3.67	12.00
fu5	Alvarez-Valdes et al. [1]	5	4	3.20	37.00
fu6	Alvarez-Valdes et al. [1]	6	5	3.33	17.89
fu7	Alvarez-Valdes et al. [1]	7	6	3.43	24.00
fu8	Alvarez-Valdes et al. [1]	8	7	3.50	24.00
fu9	Alvarez-Valdes et al. [1]	9	8	3.44	24.00
fu10	Alvarez-Valdes et al. [1]	10	9	3.50	28.00
fu	Alvarez-Valdes et al. [1]	12	11	3.42	30.00
rco1	Ribeiro et al. [10]	7	7	5.14	8.00
rco2	Ribeiro et al. [10]	14	7	5.14	15.00
santoro1	Santoro and Lemos [11]	11	11	8.00	28.00
santoro2	Santoro and Lemos [11]	11	10	4.55	6.20
santoro3	Santoro and Lemos [11]	12	8	4.33	6.50
santoro4	Santoro and Lemos [11]	17	11	4.00	19.00

Table 2: Comparison between MIP models.

Instance	NFP-CM				Improved NFP-CM			
	LB	UB	Time (s)	Nodes	LB	UB	Time (s)	Nodes
jones1	**5.00**	**5.00**	**0**	**0**	**5.00**	**5.00**	**0**	**0**
jones2	**10.00**	**10.00**	**0**	**0**	**10.00**	**10.00**	**0**	**0**
three	**6.00**	**6.00**	**0**	**0**	**6.00**	**6.00**	**0**	**0**
threep2	**9.33**	**9.33**	2	21,360	**9.33**	**9.33**	**1**	**12,963**
threep2w9	**8.00**	**8.00**	6	107,113	**8.00**	**8.00**	**1**	**15,217**
threep3	10.06	**13.53**	3,600	24,408,906	**13.53**	**13.53**	**928**	**9,711,358**
threep3w9	8.75	**11.00**	3,600	33,442,609	**11.00**	**11.00**	**1,113**	**9,036,051**
fu5	**17.89**	**17.89**	**0**	336	**17.89**	**17.89**	**0**	**235**
fu6	**23.00**	**23.00**	**0**	**428**	**23.00**	**23.00**	**0**	452
fu7	**24.00**	**24.00**	**0**	**1,324**	**24.00**	**24.00**	**0**	1,481
fu8	**24.00**	**24.00**	**0**	**851**	**24.00**	**24.00**	**0**	3,127
fu9	**25.00**	**25.00**	7	83,709	**25.00**	**25.00**	**4**	**32,911**
fu10	**28.69**	**28.69**	301	2,653,576	**28.69**	**28.69**	**33**	**253,314**
fu	28.00	**33.14**	3,600	18,761,708	**33.14**	**33.14**	**511**	**2,539,903**
rco1	**8.00**	**8.00**	**1**	14,127	**8.00**	**8.00**	**1**	**12,008**
rco2	6.67	-	3,600	17,116,136	**9.00**	-	3,600	10,534,103
santoro1	26.85	**28.00**	3,600	49,058,509	**28.00**	**28.00**	**41**	**264,666**
santoro2	**6.00**	**6.00**	474	6,958,452	**6.00**	**6.00**	**19**	**120,888**
santoro3	**6.00**	**6.00**	**5**	30,938	**6.00**	**6.00**	**5**	**17,955**
santoro4	**15.00**	**15.00**	**24**	178,258	**15.00**	**15.00**	52	**131,688**
Optimal			15				**19**	

Computational experiments show the relevance of this new set of constraints. In general, the symmetry breaking constraints improve lower bounds and reduce run-time and the number of nodes explored to prove optimality.

For further work, we intend to introduce symmetry breaking constraints for instances with non-convex pieces and study branching priorities for binary variables.

Acknowledgements

This research was supported by São Paulo Research Foundation (FAPESP), grants #2015/24987-4, #2014/23900-0, #2013/07375-0 #2012/18653-8 and #2010/10133-0 and Brazilian Federal Agency for Support and Evaluation of Graduate Education (CAPES).

References

[1] R. Alvarez-Valdes, A. Martinez, and J.M. Tamarit. A branch & bound algorithm for cutting and packing irregularly shaped pieces. *International Journal of Production Economics*, 145(2):463 – 477, 2013.

[2] Julia A. Bennell and Jose F. Oliveira. The geometry of nesting problems: A tutorial. *European Journal of Operational Research*, 184(2):397 – 415, 2008.

[3] Luiz Henrique Cherri, Leandro Resende Mundim, Marina Andretta, Franklina M. B. Toledo, José F. Oliveira, and Maria Antonia Carravilla. Mixed-integer programming models for nesting problems. *European Journal of Operational Research*, 253(3):570–583, 2016.

[4] Ahmed Elkeran. A new approach for sheet nesting problem using guided cuckoo search and pairwise clustering. *European Journal of Operational Research*, 231(3):757 – 769, 2013.

[5] Matteo Fischetti and Ivan Luzzi. Mixed-integer programming models for nesting problems. *Journal of Heuristics*, 15(3):201–226, 2009.

[6] A. M. Gomes and J. F. Oliveira. Solving irregular strip packing problems by hybridising simulated annealing and linear programming. *European Journal of Operational Research*, 171(3): 811 – 829, 2006.

[7] Donald R. Jones. A fully general, exact algorithm for nesting irregular shapes. *Journal of Global Optimization*, 59(2):367–404, 2014.

[8] Aline A.S. Leao, Franklina M.B. Toledo, José Fernando Oliveira, and Maria Antónia Carravilla. A semi-continuous mip model for the irregular strip packing problem. *International Journal of Production Research*, 54(3):712–721, 2016.

[9] B. K. Nielsen and A. Odgaard. Fast neighbourhood search for nesting problem. Technical Report 03/02, DIKU, Department of Computer Science, University of Copenhagen, 2003.

[10] Cristina Ribeiro, Maria Antónia Carravilla, and José F. Oliveira. Applying constraint logic programming to the resolution of nesting problems. *Pesquisa Operacional*, 19(2):239–247, 1999.

[11] M. C. Santoro and F. K. Lemos. Irregular packing: Milp model based on a polygonal enclosure. *Annals of Operations Research*, 235(1):693–707, 2015.

[12] F. M. B. Toledo, M. A. Carravilla, C. Ribeiro, J. F. Oliveira, and A. M. Gomes. The dotted-board model: a new mip model for nesting irregular shapes. *International Journal of Production Economics*, 145(2):478 – 487, 2013.

Aerodynamic shape optimization of an airliner elastic wing

Jan Navrátil[1,⋆]

[1]*Brno University of Technology, Faculty of Mechanical Engineering, Czech Republic*

Abstract. Aerodynamic shape optimization of a common airliner elastic wing is presented and compared with optimization of the same wing assuming rigid structure. A gradient-based optimization approach is applied in connection with adjoint method used for calculation of gradients of aerodynamic forces. The initial wing shape is parameterized by Free Form Deformation technique. The aerodynamics of the elastic wing is calculated using the fluid-structure interaction method which couples computational fluid dynamic solver with linear structural solver.

1 Introduction

An aerodynamic shape optimization using high-fidelity flow solvers has been employed for improvement of aircraft aerodynamic design over the last decades. The growing interest in this field was enabled by developments of the Computational Fluid Dynamics (CFD) solvers. CFD solvers became the accepted analysis tools in the aerospace industry reducing the number of tunnel measurements and flight tests during an aircraft development. Due to usual large number of design variables needed for aerodynamic shape design of aircraft, the gradient-based algorithms combined with adjoint solvers are only meaningful methods for practical application. Locality of those algorithms is a drawback if applied in design space where multiple local optima are likely to occur, the wing shape design is probably such a case. This restricts the process to find only a local optimum near an initial starting point. The solution might be the hybrid optimization algorithms combining non-deterministic (gradient-free) and gradient-based approaches.

The practical application of the gradient-based methods has been probably started by introduction of adjoint sensitivity analysis for Navier-Stokes equation by Pironneau in 1973 [1] and later for incompressible Euler equations [2]. The application in transonic flow regime was enabled by adjoint derivation for compressible Euler equation by Jameson in 1988 [3]. Later he extended the adjoint for Navier-Stokes equations [4]. But the stability and reliability was problem for long period of time. Today, there is only few adjoint Navier-Stockes solvers applying linearized turbulence models, others rely on approximation by frozen eddy viscosity.

Obviously, the aircraft design is multi-disciplinary problem. Increased flexibility of aircraft primary structure, as result of modern material application, requirement for lightweight structure and aerodynamically efficient shapes, even emphasizes the multi-disciplinary nature. The wing deformation due to aerodynamic load results, among others effects, in aerodynamic characteristics change. Thus, the performance gain as result of rigid model optimization might be decreased or neglected

⋆e-mail: navratil@fme.vutbr.cz

DOI: 10.1051/itmconf/20171400006

if applied to real aircraft. The solution might be inclusion of an airframe elasticity to aircraft shape optimization.

In the paper, the aerodynamic shape optimization of a common airliner elastic wing is presented and compared with optimization of the same wing assuming rigid structure. The aim is evaluation of possible benefit and the computational cost of the aerodynamic shape optimization of the elastic wing. A gradient-based optimization approach is applied in connection with adjoint method used for calculation of aerodynamic forces gradients. The coupled fluid-structure numerical solver is employed for estimation of the elastic wing aerodynamic characteristics.

2 Principle of aerodynamic shape optimization

The schematic of the applied aerodynamic shape optimization loop is given in Figure 1. An optimization algorithm directs a decision making in shape design process in order to improve desired aerodynamic characteristics (drag, glide ration, ...) by minimizing relevant objective function. A parameterization method is employed to describe a given geometry by set of parameters creating a design space. Since the parameterization deforms the surface mesh of the geometry, a mesh deformation tool must be incorporated to propagate the shape deformations into a CFD volume mesh. In the next step, the flow field is solved using CFD solver in case of a rigid model optimization, or coupled CFD with Computational Structural Mechanics (CSM) solver in case of an elastic model, providing the values of flow field variables and integral aerodynamic characteristics of the current design.

Gradients of desired variables (drag, lift, moment coefficients) with respect to all surface mesh nodes displacements are calculated on current shape using adjoint of flow equations solver. In the elastic optimization case, the gradients are calculated on current aeroelastically deformed shape. The gradients with respect to design parameters are obtained by multiplication of the surface gradient vector by parameterization Jacobian matrix. The function and gradient values are fed to the optimizer and the loop is repeated until convergence criteria are met.

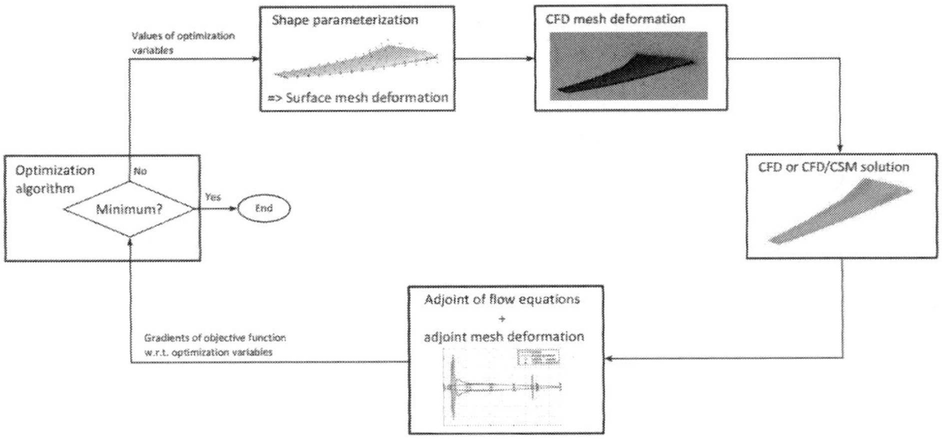

Figure 1. The aerodynamic shape optimization

3 Test cases

3.1 Tools

Flow and adjoint solvers:

The CFD code Edge [5] is used as flow and adjoint of flow equations solver. It is finite volume solver for unstructured grids which can solve 2D and 3D Euler and RANS equations, as well as the adjoint equations of the Euler and RANS (frozen viscosity) equations [6]. The time integration uses the fourth order Runge-Kutta scheme. It employs local-time-stepping, local low-speed preconditioning, multi-grid and dual-time-stepping for steady-state and time-dependent problems. The data structure of the code is edge-based. The solver can parallelize the calculation on a number of processors to solve large flow cases efficiently.

The static aeroelastic solution was obtained by coupling of the CFD Edge with the own simple CSM solver employing the beam finite elements.

Parameterization method:

A Free Form Deformation (FFD) parameterization based on NURBS was employed for a geometry parameterization. This implementation uses RBF coordinates transformation in order to better control deformations and the geometric constraints. The study of numerical properties [7] suggest the highest possible NURBS degree is beneficial from the perspective of the computational cost and the obtained optimization result.

Optimization algorithm

The optimizations were performed by gradient-based optimization algorithm - Sequential Quadratic Programming (SQP) in NLPQLP [8].

Mesh generation and deformation

The initial CFD grid was created in Ansys ICEM CFD. The spring analogy mesh deformation tool was employed to propagate surface shape changes, resulting from the optimization process, into the CFD volume mesh. For large deformations, such as the wing deformation due to the aerodynamic loading, the RBF mesh deformation combined with the spring analogy method was incorporated. The advantage of the RBF based mesh deformation is a capability to handle large deformations, but the surface deformation is not exactly recovered. The error depends on the settings of the mesh deformation solver. Therefore, the inexact surface shape was corrected by the spring analogy mesh deformation, in order to avoid introduction of an error into the optimization process by incorrect mesh deformation.

3.2 Description

The aerodynamic shape optimization test cases, based on cases proposed in [9], concern drag minimization of the transonic wing of the airliner model (so called Common Research Model [10]) at Mach number $M = 0.85$ and altitude $h = 11000m$. The lift coefficient required for steady horizontal flight in such conditions is $C_L = 0.5$. The formulation of the optimization was:

 DOI: 10.1051/itmconf/20171400006

$$\begin{aligned} \text{minimize } F(X) &= C_D \\ \text{subject to } C_L &= 0.5 \\ C_m &\geq -0.1754 \\ V &\geq V^{CRM} \\ t_f &\geq t_f^{CRM} \\ t_r &\geq t_r^{CRM} \\ &\text{fixed trailing edge} \\ &\text{wing planform shape fixed} \end{aligned} \tag{1}$$

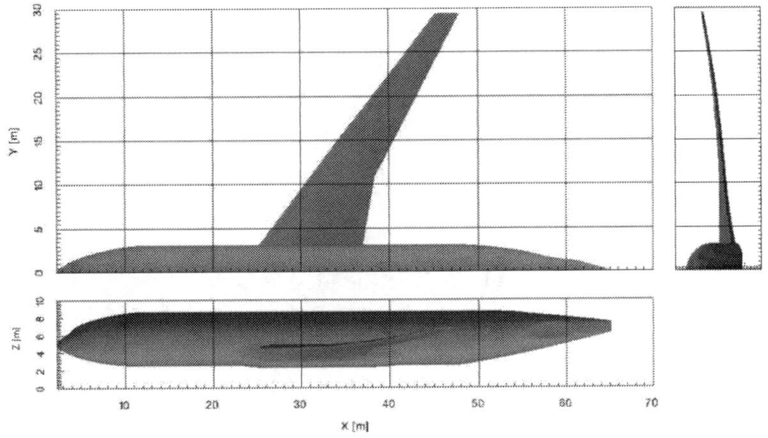

Figure 2. Common Research Model.

The goal is to decrease the drag while the lift remains constant. The geometrical constraint on the internal volume should ensure minimal space for the fuel. Other constraints on the wing thicknesses at front and rear spar positions, t_f and t_r respectively, are meant to guarantee the same minimal structural height thus the minimal structural stiffness is ensured. Therefore, the structural stiffness can remain frozen in the elastic wing optimization case.

The constraints on fixed wing planform shape and fixed trailing edge are not explicitly prescribed, but they are fulfilled by the choice of optimization variables. The Figure 3 shows the FFD lattice fitted on real geometry (in fact, the lattice is orthogonal and the geometry is transformed to fill it, description are in [7]). The blue points in the figure are optimization parameters (totally 216) allowed to move in vertical direction, therefore the airfoils shape together with twist angle are able to change while the planform shape is kept constant. The fixation of trailing edge is achieved by fixed FFD lattice points along the trailing edge, represented by red color in the figure.

Rigid wing optimization:

The initial geometry for the rigid wing optimization cases was the CRM wing as it was given in [10]. The CRM geometry was designed to provide common representative model of an airliner operating

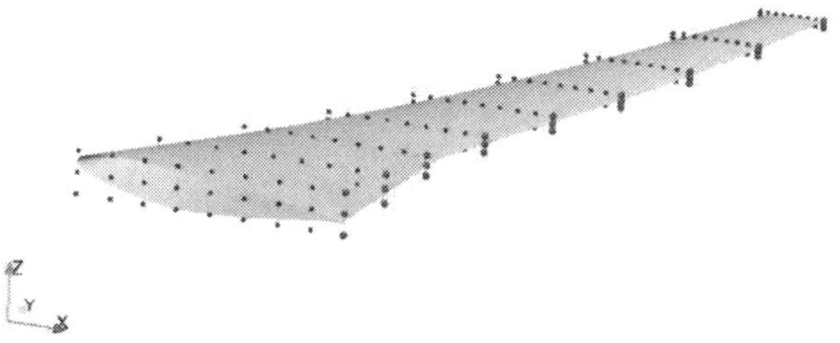

Figure 3. The FFD lattice around CRM wing, blue points are free to move in Z direction, red points are fixed.

Table 1. Aerodynamic characteristics of the initial CRM wing

	Rigid	Aeroelastic
Lift coefficient, C_L	0.5000	0.5000
Drag coefficient, C_D	0.0120	0.0120
Pitch moment coefficient, C_m	−0.1750	−0.1717
Angle of attack, α	0.721	0.816

in transonic conditions for validation of the state-of-the-art CFD solvers. Thus, the geometry of the CRM wing corresponds to flight shape (1-g shape, bended and twisted due to aerodynamic loading) at nominal cruise conditions at Mach $M = 0.85$ and $C_L = 0.5$ at altitude 12 000 m. Therefore, the subject of the optimization is a rigid wing flight shape at its nominal cruise conditions.

Elastic wing optimization:

For the elastic wing optimization, it is desirable to use an undeformed wing geometry as the static aeroelastic analysis determines correct flight shape for given operating condition. Therefore, so called jig shape was designed from the flight shape giving the undeformed wing surface and structural model. The jig shape CFD mesh was created from the mesh of the rigid wing flight shape. Applied structural model is a beam stick finite element model created by 41 beam elements. Each node of the beam element is connected with 2 additional nodes by rigid elements which are beneficial for coupling with aerodynamic surface.

The flow was solved as compressible inviscid. Assuming steady and fully turbulent flow for all designs and the constrained minimum wing wetted area, only lift-induced and wave drag can be minimized. Both flow features can be resolved by Euler flow simulation. Therefore, the optimization using the Euler flow simulation can give reliable and less expensive (in terms of computational effort) estimation of the drag reduction which could be achieved by RANS flow simulations.

 DOI: 10.1051/itmconf/20171400006

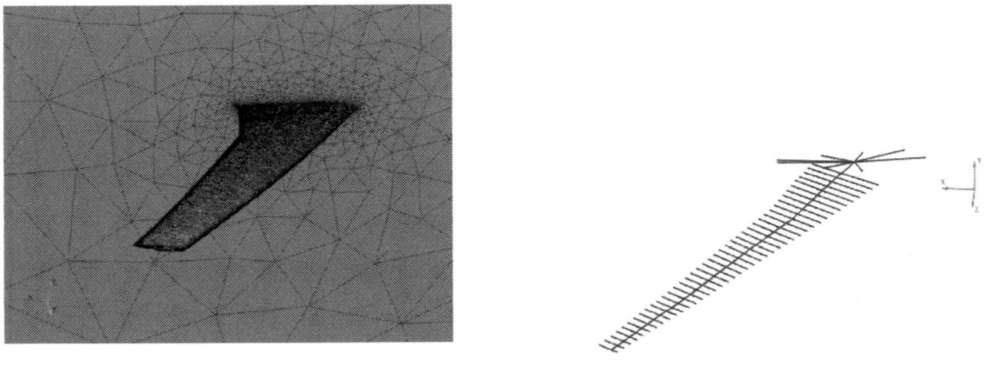

Figure 4. Computational models -the CFD mesh with 933502 nodes (left) and the beam finite element model with 41 beam elements (right).

4 Results

The comparison of both optimization approaches results is presented in Table 2. The constraint imposed on the value of the lift and the pitch moment coefficients was fulfilled, in both cases. The wing internal volume constraint was satisfied in the rigid wing case. In other case the violation was about 0.19%, what might be considered as constraint satisfaction. The optimization objective - the drag coefficient - was reduced by 6.72% and 6.17% in the rigid and elastic wing optimization, respectively. The computational cost, in terms of number of the flow and the adjoint solutions, is nearly twice as high in case of the elastic wing compared to rigid wing case.

The plots in Figure 5 suggest that constraints on the wing thickness were fulfilled in both cases. The thickness at rear spar position remained nearly unchanged in both cases, while the thickness at the front spar was slightly increased near the wing root.

The tuning of the wing shape and the twist angle of the wing sections resulted in redistribution of the lift along wing span. The resultant distribution is closer to ideal elliptical $C_L c$ distribution, as it is illustrated in Figure 6. Moreover, the lift force resultant was shifted towards wing root what might result in lower bending loading of the wing structure. The same figure shows the wing twist was changed towards higher negative values in both cases.

The Figure 8 presents the pressure coefficient distribution over the upper surface and at chosen wing sections. The plots suggest in both cases the optimization resulted in nearly shock free solution at the nominal cruise condition. The pressure change is more gradual towards the trailing edge of the optimized wing contrary to the baseline wing with steep increase of the pressure due to shock.

Table 2. Comparison of the rigid and elastic wing optimization results

	$C_{D_{opt}}$	$C_{L_{opt}}$	$C_{m_{opt}}$	V_{opt}	Cost[a]	C_D decrease
Baseline	0.0120	0.5000	−0.1750	84.46	–	–
Rigid wing	0.0112	0.5000	−0.1754	84.46	84	6.72%
Elastic wing	0.0113	0.4998	−0.1755	84.28	149	6.1739%

[a] Total cost of optimization in terms of number of flow and adjoint of flow solutions

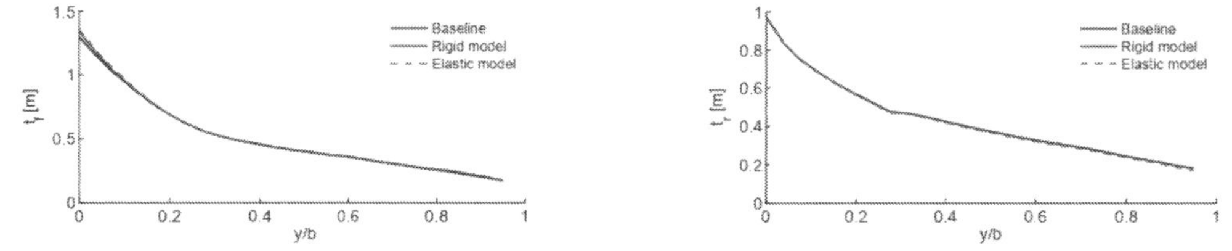

Figure 5. Comparison of the wing thickness, at position of the front (left) and rear (right) wing spar.

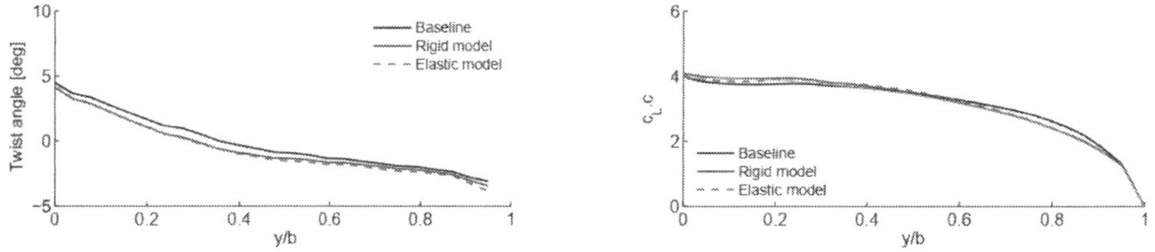

Figure 6. Comparison of span-wise wing twist and lift distributions (grey line shows the elliptical distribution of $C_L c$).

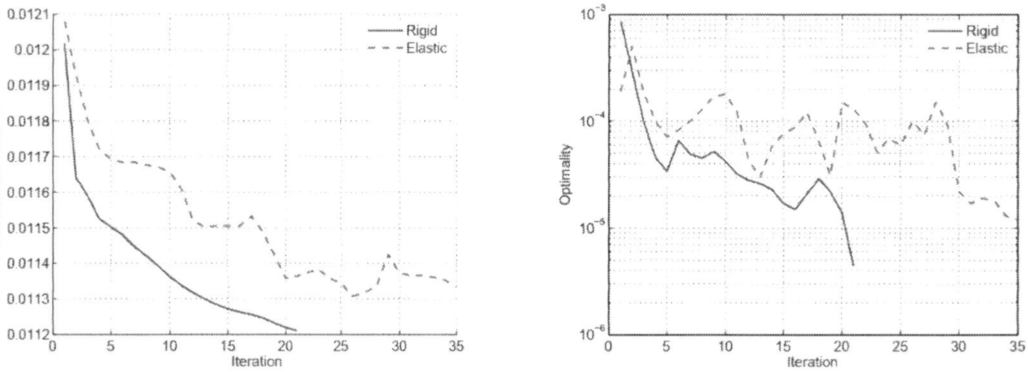

Figure 7. History of the optimization

The optimization history plots shown in Figure 7 suggest the objective function decrease was smooth and the optimization criteria were met after 21 iterations in rigid wing case. The peaks in the plot of the objective function history in case of the elastic wing together with abrupt changes of the optimality imply that there was some source of an error in the optimization chain. The probable source is the neglect of the structural deformation influence on the gradient of the aerodynamic forces and moments. Thus, implementation of the coupled fluid-structure adjoint equation solver is required.

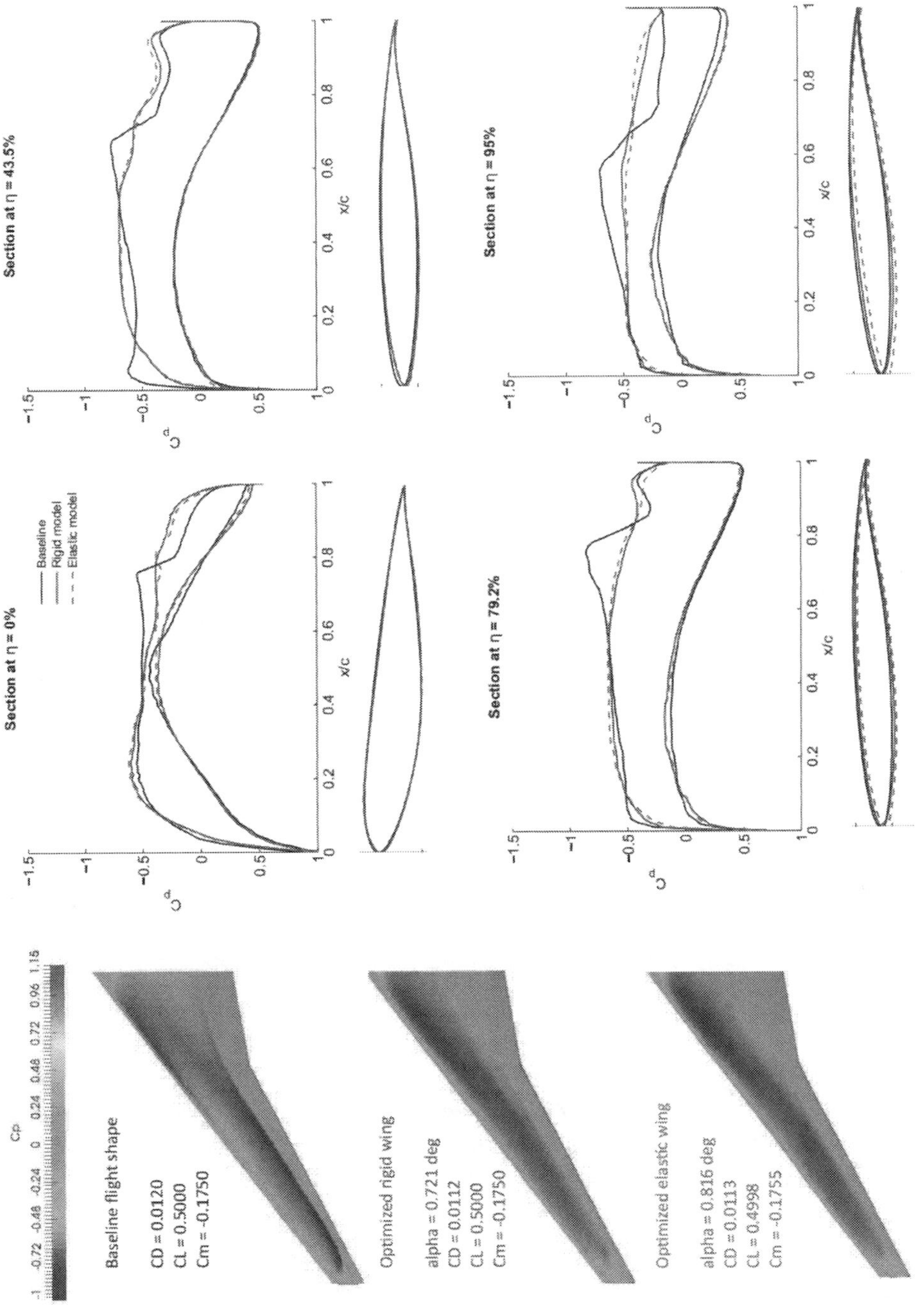

Figure 8. Comparison of surface pressure coefficient distribution - baseline and optimized wings.

DOI: 10.1051/itmconf/20171400006

5 Conclusion

The paper presents the optimization of the elastic wing compared to the rigid wing optimization. Different initial designs were applied in respective cases for the reasonable comparison. The rigid wing was optimized starting from the flight shape at the design operating conditions. Whereas in case of the elastic wing the initial design was the jig shape which under loading by aerodynamic forces at the design flight conditions deforms to the same flight shape as in rigid case.

The resultant drag reduction is nearly same in both cases, but the computational cost in the elastic wing case is almost twice as high. Moreover, the objective function decrease was not smooth as it might be expected. The several increases of the objective value during the optimization were probably result of inexact gradient calculation. Although the gradients were calculated on the aeroelastic deformed shape, the error in gradient calculation applying pure flow adjoint equation, thus neglecting the influence of the wing structure deformation on the aerodynamic forces gradients, is significant.

Thus, the derivation and implementation of the coupled fluid-structure adjoint equations is necessity for the further work on the elastic wing optimization. The real benefit of this approach is expected in the multi-point optimization considering more operation conditions.

Acknowledgments

Author would like to acknowledge funding from the Ministry of Education, Youth and Sports of the Czech Republic under the National Sustainability Programme I - Project LO1202.

References

[1] O. Pironneau, Journal of Fluid Mechanics **59**, 117 (1973)

[2] O. Pironneau, Journal of Fluid Mechanics **64**, 97 (1974)

[3] A. Jameson, Journal of scientific computing **3**, 233 (1988)

[4] A. Jameson, L. Martinelli, N. Pierce, Theoretical and computational fluid dynamics **10**, 213 (1998)

[5] P. Eliasson, *EDGE, a Navier-Stokes solver for unstructured grids*, in *Proc. to Finite Volumes for Complex Applications III* (ISBN 1 9039 9634 1, 2002), pp. 527–534

[6] O. Amoignon, M. Berggren (2003)

[7] O. Amoignon, J. Hradil, J. Navratil, *A numerical study of adaptive FFD in aerodynamic shape optimization*, in *52nd Aerospace Sciences Meeting* (2014), Vol. 899

[8] K. Schittkowski, *Nlpqlp: A fortran implementation of a sequential quadratic programming algorithm with distributed and non-monotone line search-user's guide* (2006)

[9] L. Osusky, B. H., Z.D. W., Aiaa aerodynamic design optimization discussion group (2013)

[10] J. Vassberg, M. Rivers, R. Wahls, *Development of a Common Research Model for Applied CFD Validation Studies*, in *26th AIAA Applied Aerodynamics Conference* (2008)

Resource management for flood response in Mexico

Oscar Rodriguez-Espindola[1,*], *Pavel* Albores[1], and *Christopher* Brewster[2]

[1] Aston Business School, Aston University, UK
[2] Data Science Group, Toegepast-Natuurwetenschappelijk Onderzoek (TNO), the Netherlands

Abstract. Emergency logistics is acquiring a crucial role aiming to provide adequate care and support for people affected by catastrophic situations; dealing with a large number of stakeholders, scarcity of resources and high stakes. Considering the importance of balancing resources from different participants to improve operations, this research introduces a bi-criteria response optimisation model aiming to maximise the service provided to disaster victims while making efficient use of resources. The model provides support for relief distribution, inventory management and resource allocation. But beyond the optimisation of logistical decisions, the model is able to determine which organisations to use at the appropriate timeframe, so as to prevent shortages and convergence. The model is applied to a case study on a flood in Mexico in order to compare its results with real activities performed by authorities and draw conclusions.

1 Introduction

Natural disasters are a constant threat for many countries, affecting an average of over 200 million people per year between 2000 and 2014 [1]. Emergency logistics arises as the "process of planning, managing and controlling the efficient flows of relief, information, and services from the points of origin to the points of destination to meet the urgent needs of the affected people under emergency conditions" [2].

Under disaster circumstances government agencies play a major role in aiding vulnerable population. But these are just one of the actors, being supported also by NGOs, civil organisations and aid shipped from other countries [3].

Despite the common belief that sending every resource available is the best policy to provide support, recent experiences globally have shown that it is important to manage closely resources available, to deploy the right organisations at the right time.

The purpose of this research is to provide a model capable of providing support for relief distribution and resource allocation considering the participation of different organisations.

2 Literature review

Relief distribution refers to the dispatch of different commodities to affected areas swiftly and efficiently [4]. This is a very challenging activity faced by responders [5] and also a crucial factor for high performance operations[6].

Using multi-objective approaches, Huang, Jiang [7] accounted for human suffering with three objectives; maximise lifesaving utility (i.e. preference of relief resources), minimise delay cost (similar to deprivation cost), and minimise the sum of squares of the distances between the demand fill rates and the ideal demand rate. Adıvar and Mert [8] looked at the coordination of international relief items using fuzzy logic to provide a collection-distribution plan. The model minimises cost and maximises the minimum credibility of international organisations to deliver relief items on time. Incorporating uncertainty of relief demand by using forecasting, clustering and dynamic relief supply for logistic operations related to distribution, Sheu [5] designed a model to maximise fill rate and minimise distribution cost.

Chang, Tseng [9] designed two optimisation models with different organisation levels for floods. The models cluster demand and determine local rescue bases to minimise total cost. More recently, Altay [10] developed a static model to allocate response personnel, facilities, supplies and equipment to disaster-affected jurisdictions. The alternative objective functions are to minimise total deployment time or minimise maximum response time. Including uncertainty in the state of road network, Edrissi, Poorzahedy [11] presented a formulation to coordinate agencies considering strengthening structures around vulnerable areas, retrofitting transportation links and locating and equipping emergency response centres. They introduced a dynamic version of the preparedness optimisation model to ship relief aid in the response stage.

There is only a handful of papers considering the importance of sharing resources, whether between teams [e.g. 10, 12], regions [e.g. 13], organisations [e.g. 8] or other agents [e.g. 11]. In the literature, however, there is no dynamic formulation for relief distribution optimising the number of actors involved determining the deployment and withdrawal of different agencies. Having the right expertise on the field at the right time are factors that can affect operational success in disaster response.

The purpose of this research is to provide a model incorporating several actors to identify the optimal combination of organisations and resources required per period.

3 Model design

The model developed in this research can provide support for decision-making for that interval of time. The dynamic formulation includes two objective functions, one related to cost and another focused on fill rate, thereby balancing efficiency and effectiveness. The model can determine what actors to involve and the appropriate moment for them to participate on the field.

3.1 Notation and definitions

The notation considered for the design of the model is presented as follows:

Sets

I	Supply facilities, $I = \{1, 2, 3, \ldots, I)$
J	Demand areas, $J = \{1, 2, 3, \ldots, J)$
M	Transportation modes, $M = \{1, 2, 3, \ldots, M)$
N	Products, $N = \{1, 2, 3, \ldots, N)$

O	Organisations, $O = \{1, 2, 3, \ldots, O)$
T	Periods of time, $T = \{1, 2, 3, \ldots, T)$
Parameters	
RPC	Space covered per distribution centre employee
RPS	Number of people covered per shelter employee
RPH	Number of people covered per healthcare team
APDC	Percentage of staff needed for partial opening of DCs, $0 \leq APDC \leq 1$
OND	Number of days for relief distribution per period
CP_n	Procurement cost per product n
CI_j	Storage capacity of shelter j
A_i	Capacity of DC i
$Type_j$	Type of demand area, $type_j \in [0,1]$
VOL_n	Volume per product n
WEI_n	Weight per product n
PAC_{ot}	Available personnel from organisation o for DC operation at period t
PAS_{ot}	Available personnel from organisation o for shelter operation at period t
PAH_{ot}	Available s from organisation o for healthcare at period t
PAD_{ot}	Available personnel from organisation o for distribution at period t
TRP_{ot}	Total operative staff available per organisation o at period t
G_n	Conversion factor for each product n
PRI_n	Priority of product n
SL_n	Minimum product fill rate percentage every three periods, $0 \leq SL_n \leq 1$
D_{jt}	Number of people aided in demand area j at time t
EP_k	Population to be sheltered per demand area k
F_m	Weight vehicle capacity of mode m
AVD_m	Available number of trips per day per mode m
RDP_m	Distribution personnel required per mode m
$WAGE_{ot}$	Wages paid for the activation of organisation o at period t
CS_{ijm}	Delivery cost from DC i to shelter j by mode m
IPD_{jnt}	Product demand at demand point j of product type n at period t
CON_{ijmt}	Connectivity between DC i and shelter j by mode m at period t
TV_{mot}	Number of vehicles available of type m per organisation o at period t
SA_{ijm}	Availability for distribution from i to j by mode m, $SA_{ijm} \in [0,1]$
SC_{not}	Supply capacity of product n from organisation o at period t
$DI0_{inot}$	Initial inventory of product n from organisation o at DC i at period t,
$SI0_{jnot}$	Initial inventory of product n from organisation o at j at period t
$PC0_{iot}$	Initial staff from organisation o at DC i at period t
$PS0_{jot}$	Initial staff for shelter care from organisation o at shelter j at period t
$PH0_{jot}$	Initial staff for healthcare from organisation o at shelter j at period t
$PD0_{imot}$	Initial personnel for distribution from organisation o at DC i at period t
Variables	
$DSAT_{jnt}$	Demand of product n not fulfilled on shelter j at period t
$TRAV_{ijmt}$	Number of trips deployed from DC i to shelter j by mode m at period t
NVH_{jt}	Expected number of people without healthcare at shelter j at period t
PVH_{jt}	Surplus of people covered for healthcare at shelter j at period t
NVS_{jt}	Number of people without shelter attention at shelter j at period t
PVS_{jt}	Surplus of people covered for shelter attention at shelter j at period t
APC_{iot}	Increase of DC personnel from organisation o in facility i at period t
DPC_{iot}	Decrease of DC personnel from organisation o in facility i at period t
APS_{jot}	Increase of shelter personnel from organisation o in facility j at period t

DPS_{jot}	Decrease of shelter staff from organisation o in facility j at period t
APH_{jot}	Increase of healthcare staff from organisation o in facility j at period t
DPH_{jot}	Decrease of healthcare staff from organisation o in facility j at period t
APD_{imot}	Increase of distribution staff from organisation o in i at period t
DPD_{imot}	Decrease of distribution staff from organisation o in i at period t
MAD_{jt}	Number of people with relief shortages at demand point j at period t
TSC_{jnt}	Amount of product n consumed at demand point j at period t
INV_{jnt}	Number of items of type n stored at demand point j at period t
IL_{int}	Number of items of type n stored at DC i at period t
UFR	Total unfulfilled demand
Cost	Total cost
W_{ot}	Whether to activate organisation o or not at period t
$SHIP_{ijmnt}$	Relief of type n delivered from DC i to shelter j by mode m at period t
PC_{iot}	Personnel from organisation o allocated to DC i at period t
PS_{jot}	Personnel from organisation o allocated to shelter j at period t
PH_{jot}	Staff from organisation o allocated for healthcare to shelter j at period t
PD_{imot}	Staff from organisation o to be allocated for distribution to i at period t
AV_{imot}	Vehicles of type m from organisation o allocated at DC i at period t
SUP_{inot}	Amount of product n from organisation o supplied to DC i at period t

3.2 Model formulation

The response optimisation model is looking to balance resources available during the whole disaster. The model chooses which agencies to activate depending on the circumstances and the stage of the emergency, designing a policy tailored to the situation. The formulation of the model is presented as follows:

$$MinCost = \sum_o \sum_t W_{ot} * \mathbf{WAGE_{ot}} + \sum_i \sum_n \sum_o \sum_t \mathbf{CP_n} * SUP_{inot} + \sum_i \sum_j \sum_m \sum_t \mathbf{CS_{ijm}} * TRAV_{ijmt} \quad (1)$$

$$MinUFR = \sum_j \sum_t (MAD_{jt} + NVH_{jt} + NVS_{jt}) \quad (2)$$

s.t.

$$MAD_{jt} = \frac{\sum_n DSAT_{jnt} * \mathbf{PRI_n} * \mathbf{G_n}}{\sum_n \mathbf{PRI_n}} \quad \forall j, t \quad (3)$$

$$DSAT_{jnt} = IPD_{jnt} - TSC_{jnt} \quad \forall j, n, t \quad (4)$$

$$TSC_{jnt} + TSC_{jnt+1} + TSC_{jnt+2} \geq (IPD_{jnt} + IPD_{jnt+1} + IPD_{jn} \quad \forall j, n, t \quad (5)$$

$$IL_{int} = IL_{in(t-1)} + \sum_o SUP_{inot} - \sum_j \sum_m SHIP_{ijmnt} + \sum_o DIO_i \quad \forall i, n, t \quad (6)$$

$$INV_{jnt} = INV_{jn(t-1)} + \sum_i \sum_m SHIP_{ijmnt} - TSC_{jnt} + \sum_o SIO_{jnt} \quad \forall j, n, t \quad (7)$$

$\sum_i SUP_{inot} \leq \mathbf{SC_{not}} * W_{ot}$ $\forall n,o,t$ (8)

$A_i \geq \sum_n IL_{int} * \mathbf{VOL_n}$ $\forall i,t$ (9)

$CI_j * \mathbf{TYPE_j} \geq \sum_n INV_{jnt} * \mathbf{VOL_n}$ $\forall j,t$ (10)

$\mathbf{APDC} * A_i \leq \mathbf{RPC} * \sum_o PC_{iot}$ $\forall i,t$ (11)

$D_{jt} * \mathbf{TYPE_j} \leq \mathbf{RPS} * \sum_o PS_{jot} + NVS_{jt} - PVS_{jt}$ $\forall j,t$ (12)

$D_{jt} * \mathbf{TYPE}_j \leq \mathbf{RPH} * \sum_o PH_{jot} + NVH_{jt} - PVH_{jt}$ $\forall j,t$ (13)

$PC_{iot} = PC_{io(t-1)} + APC_{iot} - DPC_{iot} + PC0_{iot}$ $\forall i,o,t$ (14)

$PS_{jot} = PS_{jo(t-1)} + APS_{jot} - DPS_{jot} + PS0_{jot}$ $\forall j,o,t$ (15)

$PH_{jot} = PH_{jo(t-1)} + APH_{jot} - DPH_{jot} + PH0_{jot}$ $\forall j,o,t$ (16)

$PD_{imot} = PD_{imo\,(t-1)} + APD_{imot} - DPD_{imot} + PD0_{imot}$ $\forall i,m,o,t$ (17)

$\sum_i PC_{iot} = \mathbf{PAC_{ot}} * W_{ot}$ $\forall o,t$ (18)

$\sum_j PS_{jot} = \mathbf{PAS_{ot}} * W_{ot}$ $\forall o,t$ (19)

$\sum_j PH_{iot} = \mathbf{PAH_{ot}} * W_{ot}$ $\forall o,t$ (20)

$\sum_i \sum_m PD_{imot} \leq \mathbf{PAD_{ot}} * W_{ot}$ $\forall o,t$ (21)

$\mathbf{TRP_{ot}} * W_{ot} \geq \sum_i PC_{iot} + \sum_j PS_{jot} + \sum_j PH_{jot} + \sum_i \sum_m PD_{im}$ $\forall o,t$ (22)

$\sum_n SHIP_{ijmnt} * \mathbf{WEI_n} \leq \mathbf{F_m} * TRAV_{ijmt} * \mathbf{CON_{ijmt}} * \mathbf{SA_{ijm}}$ $\forall i,j,m,t$ (23)

$\sum_j TRAV_{ijmt} \leq \mathbf{OND} * \sum_o AV_{imot} * \mathbf{AVD_m}$ $\forall i,m,t$ (24)

$AV_{imot} \leq \frac{PD_{imot}}{\mathbf{RDP_m}}$ $\forall i,m,o,t$ (25)

$\sum_i AV_{imot} \leq \mathbf{TV_{mot}} * W_{ot}$ $\forall m,o,t$ (26)

$IL_{inot}, MAD_{jt}, INV_{jnt}, DSAT_{jnt}, SHIP_{ijmnt}, PC_{iot}, PD_{imot}, AV_{imot}, APC_{iot}, DPC_{iot},$

$APD_{imot}, DPD_{imot,}, PH_{jot}, PS_{jot}, APH_{jot}, DPH_{jot}, APS_{jot}, DPS_{jot}, TSC_{jnt}, TRAV_{ijmt} \in Z_{\geq 0}$

$W_{ot} \in [0,1]$

Expression (1) minimises cost in terms of the number of organisations involved, procurement cost and transportation cost. Function (2) seeks the minimisation of the total unfulfillment of shelter care, healthcare and relief distribution across all demand areas and all periods. Constraint (3) determines the unfulfillment level depending on the priority of the product from the number of unsatisfied demand established on equation (4). Expression (5) is enforcing that the combination of the consumption at every demand area of every product every three periods is at least a certain percentage of the total demand at all shelters over the same periods. Equations (6) and (7) determine the levels of inventory of DCs and demand areas, respectively. Expression (8) makes sure that items procured and delivered are aligned with the supply capacity from the organisations activated. Constraints (9) and (10) restrict the storage capacity of DCs and demand areas, respectively. Expression (11) ensures a minimum number of personnel allocated to DCs even for partial opening, whereas equations (12) and (13) determine the shortage or surplus of personnel for shelter attention and healthcare. Constraints (14), (15), (16) and (17) determine the number of staff allocated for DC management, shelter care, healthcare and distribution including the rotation of personnel, whereas expressions (18), (19), (20) and (21) ensure the maximum number of staff from the activated organisations is not surpassed. Equation (22) forces the number of operative resources used per agency to be lower than the number of personnel available, including the term for healthcare teams if and only if one member is available for operative support. Expression (23) determines the number of trips from each DC to each shelter for distribution, constraint (24) establishes the number of vehicles required, equation (25) the number of employees needed, and expression (26) bounds the maximum number of vehicles available. Finally, the declaration of binary and integer variables is presented.

4 Case study: 2010 Flood at Veracruz, Mexico

4.1 Data gathering

A total of 62 Freedom of Information (FOI) were filed to regional agencies and 121 requests were made to National governmental bodies involved in the flood. The agencies involved in field operations included: DICONSA, National System of Family Development (DIF), Social Security Mexican Institute (IMSS), Civil Protection (PC), National Health Ministry (SMEXICO), State Health Ministry (SVERACRUZ), the military (SEDENA), Ministry of Interior (SEGOB) and the Navy (SEMAR).

The information gathered was in terms of relief items, prepositioned stock, shelters used, number of people sheltered per period, distribution centres activated costs, required number of employees per activity and personnel available per agency.

4.2 Solution of the model

Using the information the model was solved using the weighted-sum method and the ε-constraint method with loops of 150 iterations and a time limit of 600 seconds per iteration. A total of 142 non-dominated solutions were obtained.

More data was collected from the participating organisations. The purpose was to recreate the situation obtained according to the real activities. The only information unavailable was related to allocation of resources and services, which was optimised to obtain the *best possible outcome* that could have been achieved by authorities. The comparison can be seen in Figure 1.

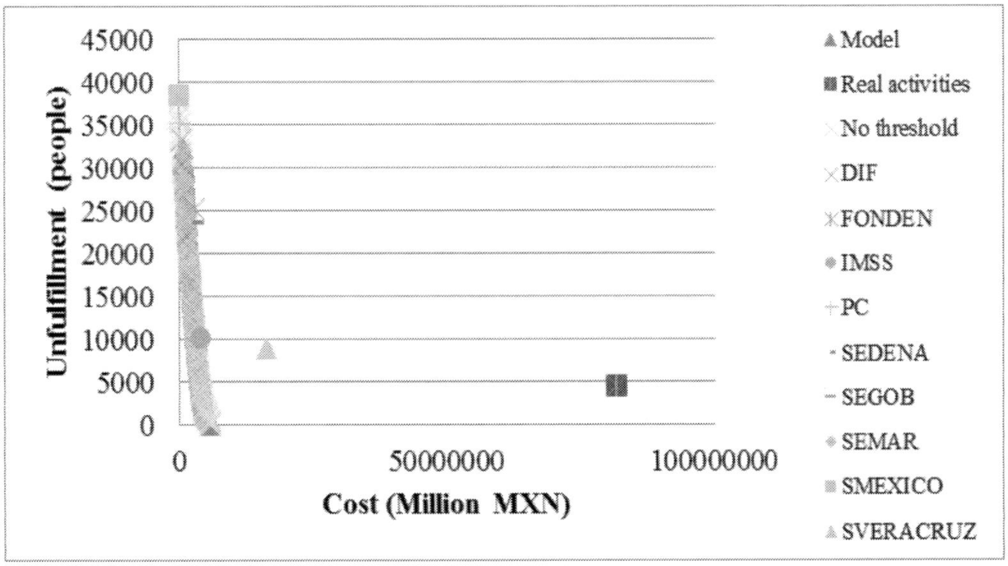

Fig. 1. Comparison of the model to independent decision-making and real activities

The figure shows how independent decision-making was not a suitable policy for the disaster as there is no organisation capable of coping with the conditions of the disaster. It can also be seen that assuming a coordinated approach the real decisions made by authorities represented an acceptable fill rate but at a very high cost compared to the potential solution delivered by the model.

5 Conclusions

This research introduced a novel dynamic optimisation model for disaster management considering the involvement of several organisations. Resource management is a significant component of response activities such as relief distribution and inventory management. The results of the model showed the potential to handle several organisations and deploy them according to the circumstances. The model was capable of providing a 100% of satisfaction rate with the resources available without the need to involve all of the organisations that participated in reality in the disaster.

The comparison with independent decision-making showed the incapability of any of the organisations involved to handle the situation on their own, whereas the comparison to real activities exhibited the room of improvement for future operations in terms of cost and fill rate.

The result of the study highlights the importance of considering the management of resources for logistical decisions in disaster management and the need to integrate the different actors to avoid duplications of effort and achieve more successful operations.

6 References

1. CRED, Advanced Search. 2015, Université Catholique de Louvain: Brussels, Belgium.
2. J.B. Sheu, Transp. Res. Part E: Log. and Transp. Rev. 43, 6, 655-659 (2007)
3. I.M. Nolte, E.C. Martin, S. Boenigk, Pub. Man. Rev. 14, 6, 707-730 (2012)
4. L. Özdamar, E. Ekinci, B. Küçükyazici, Annals of Oper. Res. 129, 1, 217-245 (2004)
5. J.B. Sheu, Transp. Res. Part E: Log. and Transp. Rev. 2007. 43, 6, 687-709 (2007)

6. L.E. De la Torre, I.S. Dolinskaya, K.R. Smilowitz, Soc. Econ. Plan. Sci. 46, 1, 88-97 (2012)
7. K. Huang, Y. Jiang, Y. Yuan, L. Zhao, Transp. Res. Part E: Log. and Transp. Rev, 75, 1-17 (2015)
8. B. Adıvar, A. Mert, Fuzzy Opt. & Dec. Making, 9, 4, 413-433 (2010)
9. M.S. Chang, Y.L. Tseng, J.W. Chen, Transp. Res. Part E: Log. and Transp. Rev. 43, 6, 737-754 (2007)
10. N. Altay, IMA J. of Man. Math. 24, 2, 253-266 (2013)
11. Edrissi, H. Poorzahedy, H. Nassiri, M. Nourinejad, Eur. J. of Op. Res. 229, 1, 261-275, (2013)
12. G.G. Brown, A.L. Vassiliou, Nav. Res. Log. 40, 1, 1-23 (1993)
13. H. Arora, T.S. Raghu, A. Vinze, Dec. Sup. Syst. 50, 1, 304-315 (2010)

ITM Web of Conferences **14**, 00008 (2017) DOI: 10.1051/itmconf/20171400008
APMOD 2016

Terminal Control Area Aircraft Scheduling and Trajectory Optimization Approaches

Marcella Samà[1,*] [**], *Konstantin* Palagachev[2], *Andrea* D'Ariano[1], *Matthias* Gerdts[2], and *Dario* Pacciarelli[1]

[1] *Roma Tre University, Department of Engineering, Section of Computer Science and Automation, Via della Vasca Navale, 79, 00146 Rome, Italy*

[2] *Universität der Bundeswehr München, Fakultät für Luft- und Raumfahrttechnik, Institut für Mathematik und Rechneranwendung, Werner-Heisenberg-Weg 39, D-85577 Neubiberg/München, Germany*

Abstract. Aviation authorities are seeking optimization methods to better use the available infrastructure and better manage aircraft movements. This paper deals with the real-time scheduling of take-off and landing aircraft at a busy terminal control area and with the optimization of aircraft trajectories during the landing procedures. The first problem aims to reduce the propagation of delays, while the second problem aims to either minimize the travel time or reduce the fuel consumption. Both problems are particularly complex, since the first one is NP-hard while the second one is nonlinear and a combined solution needs to be computed in a short-time during operations. This paper proposes a framework for the lexicographic optimization of the two problems. Computational experiments are performed for the Milano Malpensa airport and show the existing gaps between the performance indicators of the two problems when different lexicographic optimization approaches are considered.

1 Introduction

Air traffic control must ensure safe, ordered and rapid transit of aircraft on the ground and in the air segments. Airports have become the major bottleneck in air traffic control due to the strong increase in air traffic demand of the last decades. Operations in air traffic control are performed by human controllers aided by some computer support, often limited to a graphical representation of the aircraft position and speed. The controllers of the Terminal Control Area (TCA) determine feasible plans by selecting the order in which the different aircraft perform their operations and the aircraft timing in each airport resource. Usually the controllers start from a solution computed using some scheduling rule, e.g. the First Come First Served (FCFS) rule, and make then adjustments to improve the schedule. However, the resulting solution does not fully take into consideration neither the propagation of aircraft delays nor the minimization of other relevant performance indicators [16].

*e-mail: sama@ing.uniroma3.it
**Acknowledgement: The corresponding author was supported by the German academic exchange service (DAAD) with a Research Grant for PhD candidates, Funding Program Number 57130097

This paper deals with the Terminal Control Area Aircraft Scheduling Problem (TCA-ASP) and the Aircraft Trajectory Optimization Problem (ATOP) for landing operations in a busy TCA. The first problem is to compute a conflict-free schedule for all aircraft with the minimization of the overall aircraft delays. The second problem is to compute a landing trajectory for each aircraft with either the minimization of the final time or the minimization of the fuel consumption. The trajectory with minimal travel time helps to reduce the individual aircraft delay, while trajectories with minimal fuel consumption can increase aircraft delays. According to recent surveys on the aircraft scheduling literature [1–3, 8, 11], there is still a need to develop integrated optimization approaches for solving the TCA-ASP and the ATOP. This topic requires to take simultaneously into consideration the following aspects:

1. In the TCA-ASP, the optimization approach should be able to incorporate all the information that ensures that the aircraft schedule is compliant with the safety regulations. In most of the existing approaches, the characteristics of the airport infrastructure are drastically simplified and the flight paths are aggregated. Thus potential aircraft conflicts may not be detected and solved at the level of runways, ground and air segments of the TCA.

2. In the ATOP, the trajectory optimization approach should be done in order to ensure safety and efficiency of the flight. A detailed approach for planning the flying time of each aircraft should take into account the behaviour of all aircrafts that share the same TCA resources.

3. The available time for developing a new schedule for the integrated problem can be very limited, since a computerized scheduler should be able to promptly react to any change occurring in the TCA.

This paper addresses the first aspect by formulating the TCA-ASP as a job shop scheduling model with additional constraints. This modelling approach is able to consider the airspace interactions between aircraft in each TCA resource. However, a pre-defined trajectory has to be modelled for each aircraft. The trajectory can be fine-tuned before/after solving the TCA-ASP.

The second aspect is addressed by formulating and solving an optimal control problem, whose solution provides the reference trajectory for each aircraft flying through each resource.

The third aspect requires to use fast and effective approaches to solve the TCA-ASP and the ATOP, and for integrating their solutions. In this paper, we solve the TCA-ASP via *AGLIBRARY*, a dedicated software developed by Roma Tre University, and we compute a near-optimal solution to the ATOP of each landing aircraft via *OCPID-DAE1*, a dedicated software developed by the University of the Federal Armed Forces at Munich. The integration of the TCA-ASP and ATOP is obtained via a new optimization framework developed in this paper.

The remaining of this paper is organized as follows. Section 2 gives a brief description of the TCA-ASP models and algorithms of AGLIBRARY, Section 3 describes the approach used in OCPID-DAE1 to solve the ATOP, Section 4 presents the new optimization framework for integrating the TCA-ASP and ATOP solutions. The integrated solution is obtained by a lexicographic optimization, in which the TCA-ASP and ATOP are solved sequentially. We study various settings of the framework in which we change the order in which the two optimization problems are solved and the performance indicators to be optimized. Section 5 concludes the paper and outlines further research directions.

2 Terminal Area Aircraft Scheduling Approach

Let us consider a set of landing and take-off aircraft to be scheduled in the TCA. For each aircraft, a feasible trajectory in the TCA is provided together with its current position, and its scheduled occupancy times in the required TCA resources to accomplish the arriving/departing procedures. The

TCA-ASP is to assign the start time to each aircraft in each TCA resource it crosses in such a way that all the potential conflict situations between aircraft are solved, the safety constraints are respected and the propagation of aircraft delays is minimized.

To model the TCA-ASP, we follow the approach of Bianco et al. [4], based on the no-wait version of the job shop scheduling problem. However, this paper is based on the alternative graph model introduced by Mascis and Pacciarelli [9]. The alternative graph model is able to model with higher precision some relevant TCA-ASP aspects such as modeling the airborne holding, the aircraft trajectory in each TCA resource, the hosting of multiple aircraft simultaneously in each air segment, and the hosting of individual aircraft simultaneously in each runway.

We now formalize TCA-ASP model. An *operation* i denotes the traversing of a TCA resource (i.e. holding circle, air segment, runway) by a job (i.e. aircraft). The timing of an operation i represents the aircraft trajectory in the corresponding resource. Given the timing related to each aircraft operation, the associated schedule is *conflict-free* if the minimum time separation constraints are satisfied on each resource.

Let $G = (N, F, A)$ be the alternative graph composed of the following sets: $N = \{0, 1, ..., n\}$ is the set of *nodes*, where nodes 0 and n represent the start and the end operations of the schedule, while the other nodes are related to the start of the other operations; F is the set of *fixed directed arcs* that model the sequence of operations to be executed by each aircraft; A is the set of *alternative pairs* that model the sequencing decision and inter-aircraft safety separations. Each pair is composed of two alternative directed arcs.

Let t_i be the start time of operation i, each node $i \in N$ of the alternative graph is associated with t_i, and corresponds to the entrance of the associated aircraft in the associated resource. By definition, the start time of the schedule is a known value, e.g. $t_0 = 0$, and the end time of the schedule is a variable t_n.

Let $\sigma(i)$ denote the operation following i. Each fixed directed arc $(i, \sigma(i)) \in F$ $((\sigma(i), i) \in F)$ has a weight $w^F_{i\sigma(i)}$ $(w^F_{\sigma(i)i})$, i.e. the constraint $t_{\sigma(i)} \geq t_i + w^F_{i\sigma(i)}$ $(t_i \geq t_{\sigma(i)} + w^F_{\sigma(i)i})$. The fixed arc weight $w^F_{i\sigma(i)}$ $(-w^F_{\sigma(i)i})$ models the minimum (maximum) traversing time between the start of i and the start of $\sigma(i)$.

Each alternative pair $((i, j), (h, k)) \in A$ has two arcs with weight w^A_{ij} and w^A_{hk}. The alternative arc weight w^A_{ij} represents a minimum separation time between the start of i and the start j. In particular, w^A_{ij} (w^A_{hk}) can be sequence-dependent, when nodes i and j (h and k) are operations of different jobs. Also, there can be multiple alternative arcs between nodes i and j.

A *selection* S is a set of alternative arcs, at most one from each pair. Given the alternative pair $((i, j), (h, k)) \in A$, we select either the arc (i, j) or the arc (h, k), i.e. either the constraint $t_j \geq t_i + w^A_{ij}$ or the constraint $t_k \geq t_h + w^A_{hk}$.

An ASP solution is a *complete* selection S^c, where an arc for each alternative pair of the set A is selected and the connected graph (N, F, S^c) has no positive weight cycles. Note that a positive weight cycle represents an operation preceding itself, which is an *infeasibility*. Given a feasible schedule S^c, t_i is the length of a longest path from 0 to i $(l^{S^c}(0, i))$. A fixed arc (k, n) between the end node k of each job and node n is added to the alternative graph, with weight equal to the scheduled arrival time at the runway with negative sign. A selection S^c is optimal if $l^{S^c}(0, n)$ is minimum over all the solutions. In this work, we minimize the maximum tardiness, that is $l^{S^c}(0, n)$.

We compute a near-optimal solution to the proposed TCA-ASP formulation via a branch & bound algorithm [6] truncated after a maximum computation time (i.e. after 180 seconds), implemented in the AGLIBRARY solver. A more detailed description of the alternative graph models, algorithms and solver can be found in D'Ariano et al. [5, 6] and Samà et al. [13–15].

3 Aircraft Trajectory Optimization Approach

The ATOP is an optimal control problem, representing a simplified, two-dimensional point mass model of the aircraft's dynamics. It includes the state variables x, h, V and γ, representing respectively the aircraft's position, altitude, speed and climb angle. The control inputs to the model are given by the angle of attack α and the thrust level position δ_T, constrained in the interval $[0, 1]$. Hence the actual thrust applied to the aircraft is given by $T = T_{max}\delta_T$, where T_{max} is the maximum thrust available. The list L and drag D forces, acting on the aircraft, are given by

$$L = \frac{\rho}{2} \cdot V^2 \cdot S_{ref} \cdot C_L \qquad \textit{and} \qquad D = \frac{\rho}{2} \cdot V^2 \cdot S_{ref} \cdot C_D$$

where C_L and C_D are respectively the lift and drag coefficients

$$C_L = C_{L_0} + C_{L_\alpha}\alpha \qquad \textit{and} \qquad C_D = C_{D_0} + C_{D_2}C_L^2.$$

Note that C_{L_0}, C_{L_α}, C_{D_0} and C_{D_2} are unit-less parameters, depending on the flat's configuration of the aircraft, as well as on the aircraft itself. Furthermore, S_{ref} is the aircraft's wing reference surface, m is the aircraft's mass, ρ is the air density, while g is the gravitational acceleration. Hence ATOP reads as follow

$$\textit{Minimize} \qquad \omega \cdot t_f + (1-\omega)\int_0^{t_f} \|\delta_T(t)\|^2 dt \tag{1}$$

$$\textit{subject to} \qquad \dot{x}(t) = V(t) \cdot \cos\gamma(t) \tag{2}$$

$$\dot{h}(t) = V(t) \cdot \sin\gamma(t) \tag{3}$$

$$\dot{V}(t) = \frac{T(t) - D(t)}{m} - g \cdot \sin\gamma(t) \tag{4}$$

$$\dot{\gamma}(t) = \frac{L(t) - m \cdot g \cdot \cos\gamma(t)}{m \cdot V(t)} \tag{5}$$

$$(x, h, V, \gamma)(0) = (x_0, h_0, V_0, \gamma_0) \tag{6}$$

$$(x, h, V, \gamma)(t_f) = (x_f, h_f, V_f, \gamma_f). \tag{7}$$

Note that in (1) the parameter ω is responsible for the minimum travel time problem ($\omega = 1$, ATOP-tt), the minimum fuel consumption problem ($\omega = 0$, ATOP-fc) or any combination of the two problems ($\omega \in (0, 1)$). Furthermore, equations (6)-(7) are responsible for fixing the states of the problem at initial and final time (depending on the resource through which the aircraft is flying). We emphasize that for sake of simplicity, we are not including any bounds for the state variables, as well as any constraint in the formulation of ATOP. For more precise formulation, we refer the interested reader to [10, 12].

In order to compute the solution of ATOP, we adopt a standard technique in optimal control theory, known as discretize-then-optimize. It consists in the approximation of the original infinite dimensional problem by a suitable finite dimensional one. Then, the approximated problem is solved and the solution is interpreted as an approximation of the solution of the original problem (see [7] for more details). One of the main advantages is that the optimal solution of ATOP can be reached with high precision and its optimality is guaranteed by mathematical theory. Furthermore, the optimal solution is easily recomputable, once the dynamics or the cost function are perturbed.

4 Integrated Approaches

The problems studied in the previous sections are now combined in order to find an integrated solution. To this aim, we use the framework of Figure 1.

 DOI: 10.1051/itmconf/20171400008

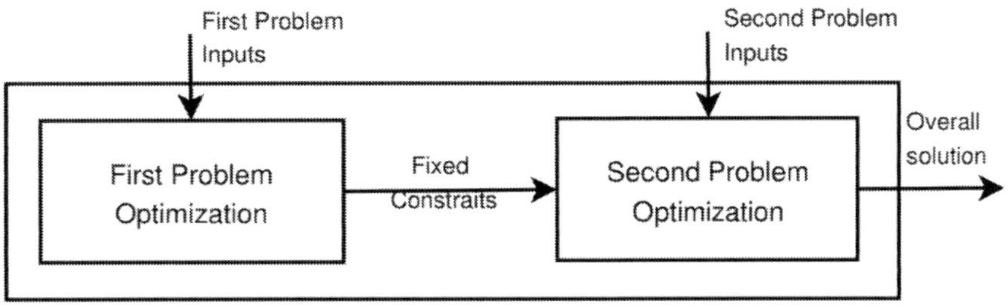

Figure 1. Solution framework

The idea is to solve the two problems sequentially and to define a lexicographic order of importance for the performance indicators to be optimized. Therefore, the following alternative integrated approaches are investigated:

- First ATOP, then TCA-ASP: We first compute the near-optimal trajectory of each landing aircraft in each TCA resource it requires. This trajectory is found by either minimizing the aircraft travel time (ATOP-tt) or the fuel consumption (ATOP-fc), and it is used to form the job in the alternative graph corresponding to the investigated aircraft. In this integrated approach, we need to set a fixed travel time for the aircraft in the TCA-ASP problem. This is modelled by setting the weight $w^F_{i\sigma(i)}$ of each fixed arc $(i, \sigma(i))$ equal to the corresponding traversing time in the ATOP, and $w^F_{\sigma(i)i} = -w^F_{i\sigma(i)}$.
- First TCA-ASP, then ATOP: We first compute the near-optimal solution to the TCA-ASP. The resulting connected graph (N, F, S^c) is then studied in order to find a range of feasible values for the travel time of each landing aircraft in each resource. Specifically, we set a minimum and a maximum value of the travel time such that the selection S^c is respected in any feasible ATOP solution. Clearly, the ATOP optimization only focuses on the aircraft that offer some degree of flexibility, i.e. on the aircraft having a different value of the minimum and maximum travel times on at least one TCA resource.

5 Computational Experiments

Figure 2 shows the TCA scheme of Milano Malpensa airport (MXP). There are two interdependent runways (RWY 35L, RWY 35R), used both for departing and arriving procedures. The MXP resources are 3 airborne holding circles (resources 1-3, named TOR, MBR, SRN), 11 air segments for arriving procedures (resources 4-14), 1 common glide path (resource 15), and 2 runways (resources 16-17).

Table 1 describes the 15 instances that we generated with random entrance delays. Column 1 presents the TCA (FCO and MXP), Columns 2-3 the number of landing and take-off aircraft, Columns 4-5 the maximum and average entrance delays (in seconds). The entrance delays are randomly generated for the first half aircraft entering the TCA according to a given distribution (either negative exponential or Gaussian distributions). The exit delays are measured as a positive deviation from the scheduled take-off/landing time.

Table 1. Problem instances

TCA	Landing Aircraft	Take-off Aircraft	Max Entrance Delay (sec)	Avg Entrance Delay (sec)
MXP	14	6	1792.9	869.6

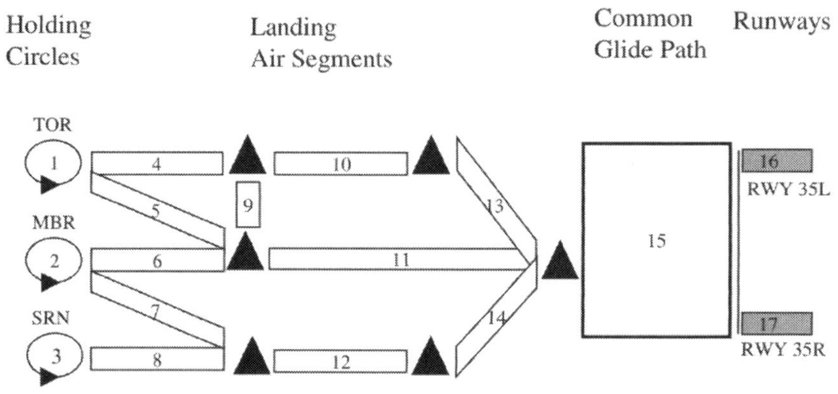

Figure 2. Milano Malpensa (MXP) Terminal Control Area

Preliminary computational experiments have been performed on a Quad-Core Intel Xeon E5 3.7GHz processor with 32 GB RAM, under OS X 10.10.3. Table 2 presents an average comparison of the various integrated approaches on the 15 instances of Table 1. Column 1 indicates the integrated approach configuration and Columns 2-4 the average value of a specific performance indicator. The best value for each column is reported in bold. The results show an interesting trade-off between the performance indicators related to the different problems. When minimizing the maximum tardiness and the travel time, the lexicographic order in which the two problems are tackled does not impact on the solution quality, suggesting a strong correlation between the two indicators. A different trend is observed when dealing with the minimization of the fuel consumption. The lexicographic order in which the two problems are tackled impacts the solution quality in terms of all performance indicators.

Table 2. Trade-off between the three performance indicators

Integrated Approach	Max Tardiness	Travel Time	Fuel Consumption
TCA-ASP/ATOP-tt	**1494.3**	**9812.0**	2369.9
TCA-ASP/ATOP-fc	**1494.3**	12188.8	756.4
ATOP-tt/TCA-ASP	**1494.3**	**9812.0**	2369.9
ATOP-fc/TCA-ASP	1729.2	13278.0	**276.0**

6 Conclusions and Further Research

This paper addresses the problem of developing an integrated framework for the TCA-ASP and the ATOP. Various integrated approaches are investigated and tested on a practical case study. Preliminary results quantify the gap between the performance indicators, highlighting the multi-objective nature of the problem.

Further research should be dedicated at least to the following issues: evaluate the alternative framework on different instances, improve the effectiveness of the integrated approaches, validate the approaches at the traffic control center.

References

[1] Allahverdi, A., Ng, C.T., Cheng, T.C.E., Kovalyov, M.Y. (2008) A survey of scheduling problems with setup times or costs. *European Journal of Operational Research* **187 (3)** 985–1032

[2] Ball, M.O., Barnhart, C., Nemhauser, G., Odoni, A. (2007) Air Transportation: Irregular Operations and Control. In: G. Laporte and C. Barnhart (Eds.). *Handbooks in Operations Research and Management Science* **14** 1–67

[3] Bennell, J.A., Mesgarpour, M., Potts, C.N. (2011) Airport runway scheduling. *4OR – Quarterly Journal of Operations Research* **9 (2)** 115–138

[4] Bianco, L., Dell'Olmo, P., Giordani, S. (2006) Scheduling models for air traffic control in terminal areas. *Journal of Scheduling* **9 (3)** 180–197

[5] D'Ariano, A., Pistelli, M., Pacciarelli, D. (2012) Aircraft retiming and rerouting in vicinity of airports. *IET Intelligent Transport Systems* **6 (4)** 433–443

[6] D'Ariano, A., Pacciarelli, D., Pistelli, M., Pranzo, M. (2015) Real-time scheduling of aircraft arrivals and departures in a terminal maneuvering area. *Networks* **65 (3)** 212–227

[7] Gerdts, M. (2012) *Optimal Control of ODE and DAEs*, 458p. Walter de Gruyter GmbH & Co.KG, Berlin Boston.

[8] Kuchar, J.K., Yang, L.C. (2000) A Review of Conflict Detection and Resolution Modeling Methods. *IEEE Trans. on Intelligent Transp. Systems* **4 (1)** 179–189

[9] Mascis, A., Pacciarelli, D. (2002) Job shop scheduling with blocking and no-wait constraints. *European Journal of Operational Research* **143** (3), 498–517

[10] Palagachev, K., Rieck, M., Gerdts, M. (2013) A Mixed-Integer Optimal Control Approach for Aircraft Landing Model. *Proceedings of the 3rd International Conference on Models and Technology for Intelligent Transportation Systems (MT-ITS 2013)*, 10 p., Dresden, Germany

[11] Pellegrini, P., Rodriguez, J. (2013) Single European sky and single European railway area: A system level analysis of air and rail transportation. *Transportation Research Part A* **57 (1)** 64–86

[12] Rieck, M., Richter, M., Holzapfel, F. (2013) Discrete Control Dependent Constraints in Multiple Shooting Optimal Control Problems. *Proceedings of the AIAA Guidance, Navigation, and Control (GNC) Conference, Guidance, Navigation, and Control and Co-located Conferences (AIAA 2013-4526)*

[13] Samà, M., D'Ariano, A., Pacciarelli, D. (2013) Rolling Horizon Approach for Aircraft Scheduling in the Terminal Control Area of Busy Airports. *Transportation Research Part E* **60** (1), 140–155

[14] Samà, M., D'Ariano, A., D'Ariano, P. Pacciarelli, D. (2014) Optimal aircraft scheduling and routing at a terminal control area during disturbances. *Transportation Research Part C* **47 (1)** 61–85

[15] Samà, M., D'Ariano, A., D'Ariano, P. Pacciarelli, D. (2015) Air traffic optimization models for aircraft delay and travel time minimization in terminal control areas *Public Transport* **7 (3)** 321–337

[16] Samà, M., D'Ariano, A., D'Ariano, P. Pacciarelli, D. (2016) Scheduling models for optimal aircraft traffic control at busy airports: Tardiness, priorities, equity and violations considerations *OMEGA* doi:10.1016/j.omega.2016.04.003

Shaking the trees: Abilities and Capabilities of Regression and Decision Trees for Political Science

Christoph Waldhauser[1,⋆] and *Ronald* Hochreiter[2,⋆⋆]

[1]*KDSS Data Science, 1060 Vienna, Austria.*
[2]*WU Vienna University of Economics and Business, Welthandelsplatz 1, 1020 Vienna, Austria.*

Abstract. When committing to quantitative political science, a researcher has a wealth of methods to choose from. In this paper we compare the established method of analyzing roll call data using W-NOMINATE scores to a data-driven supervised machine learning method: Regression and Decision Trees (RDTs). To do this, we defined two scenarios, one pertaining to an analytical goal, the other being aimed at predicting unknown voting behavior. The suitability of both methods is measured in the dimensions of consistency, tolerance towards misspecification, prediction quality and overall variability. We find that RDTs are at least as suitable as the established method, at lower computational expense and are more forgiving with respect to misspecification.

1 Introduction

Roll call voting behavior of legislators in the US Congress is at the focal point of many political science research efforts. A traditional model relies on W-NOMINATE scores to condense the many roll call records into a small number of dimensions. We offer a different approach, using a supervised machine learning method that is of equal potency. We test this using two different scenarios, one pertaining to analytical questions of where a vote can be located at in the political sphere. The other scenario relates to predicting how the other legislators will vote, using only a small number of "declared" or "known" legislators.

This paper is organized as follows. First general observations regarding the suitability of machine learning methods in political science are being discussed. There, we focus on both the popularity of these methods in general and the contrasted methods in particular, as well as on caveats specific to applications in voting studies. We then turn the reader's attention on the methods compared in this paper. We put special emphasis on using them in both predictive and analytic roles.

We then introduce the testing methodology to set a fixed set of scenarios in which both methods are to be used and explain the indicators employed to ascertain a method's suitability. The next section describes the results obtained from applying the methods to the scenarios outlined before. Finally, we discuss these results and offer some concluding remarks.

⋆e-mail: chw@kdss.at
⋆⋆e-mail: ronald.hochreiter@wu.ac.at

 DOI: 10.1051/itmconf/20171400009

2 Machine Learning and Voting Studies

As outlined above, this paper is about applying a premier machine learning algorithm to the study of voting behavior in the US House of Representatives. These applications have a rather long history in engineering and computer science, as they are considered difficult problems. This history is so profound, that there is even a voting data set contained in the wealth of the UCI machine learning repository: the 1984 Congressional Votes Records Database [1].

That data set contains the voting behavior of all the members of the 1984 US House of Representatives on key decisions of that year. It has been used numerous times over the years in different levels of applications ranging from articles [2–5] to books [6] to PhD theses [7]. Most of these works exhibit some degree of neglect towards the specialties [8, 9] of a political science domain, some even grave mistakes. While this negligence is of little importance in evaluating the quality of machine learning approaches or algorithms, it is crucial for assessing a method's usability in political science. It is only after the peculiarities of the political science domain have been taken into consideration and molded into an algorithm, that the practitioner can successfully wield the tool.

The 1984 Congressional Voting Records Database is a data set that details the voting behavior of all the Representatives in the US House of Representatives of the year 1984. The items that are listed, are 16 key votes, as they were singled out by the Congressional Quarterly Almanac of 1985 [10]. While the CQA identifies nine different voting behaviors, in the data set, the variables are recoded into three-valued categorical variables. Additionally, party membership of the Representatives is coded as a dichotomous variable. The Representatives themselves are ordered by state (and alphabetically thereafter), thus generating an inherent structure to the data set, that is neither described nor accounted for. While [11, p. 85] argue that 1984 was a pretty normal year regarding party influence on the Representatives, the specifics of the data set make it rather cumbersome to use in the evaluation of party affiliation prediction algorithms. We will now examine the conceptual shortcomings of the data set in detail.

Representatives in the US Congress can take different actions when a bill hits the floor. Obviously, they can vote yea and nay. They can also declare how they would have voted, if they had been present. And they can abstain from a vote.

> "The CQA lists nine different types of votes: voted for, paired for, and announced for (these three simplified to yea), voted against, paired against, and announced against (these three simplified to nay), voted present, voted present to avoid conflict of interest, and did not vote or otherwise make a position known (these three simplified to an unknown disposition)."(from the data set description as provided by [1]).

These simplifications in the data set are not uncontested. The CQA print version, for instance does not follow this simplification, but rather just denotes the pure Yeas and Nays. While the details of the nature of the vote may have some significance for democratic transparency, it does not change the outcome of a vote. If a majority votes Yea, announces or pairs for a motion, the motion is being accepted. For the pure assessment of the outcome of a vote a detailed modeling of the vote's nature is not necessary. However, if it is not only the outcome that is interesting, but the voting behavior itself, a lot of information is discarded by following the simplifications of the data set.

2.1 Voting Data considerations regarding missing values

Voting behavior data, or more precisely, roll call votes of the US Congress as used here, are records of votes cast by legislators. These votes can be either in the affirmative of or against a proposed piece of legislation. However, there is a third kind of vote besides the more expressive Yea and Nay:

abstention. One approach is to consider abstentions as some sort of missing value: a legislator that did not find the time to cast a vote. We shall now consider the effects of assuming abstentions to be simply missing values.

Missing values are instances of a variable that do not have a reliable data point associated with them. The traditional text-book example would be a missing response from a questionnaire, or an obviously failing gauge in a more technical setting. The treatment of traditional missing values has been discussed in length in literature, [5], as a more recent example, suggest one of the following methods to deal with missing values:

> "A simple solution involves ignoring instances and/or attributes containing missing values, but the waste of data may be considerable and incomplete data sets may lead to biased statistical analyses. Another alternative is to substitute the missing values by a constant. However, it assumes that all missing values represent the same value, leading to considerable distortions. The substitution by the mean/mode value is common and sometimes can even lead to reasonable results. However, this approach does not take into account the between-attribute relationships, which are usually explored by data mining methods. Therefore, a more interesting approach involves trying to fill missing values by preserving such relationships."[5, pp. 245]

They continue detailing their approach, and evaluate it using the CQA data set. There, they count "435 instances, of which 203 have missing values. These instances were removed (to make the prediction evaluation possible) and the proposed method was employed in the remaining ones"[5, p. 251]. The authors thus did exactly what they were warned not to do in the data set description. While it is certainly possible to think of not cast votes as simply missing values, because the Representatives did either not have an opinion on a subject, or did not find the time to come to Congress, this model appears presumptuous if applied to key votes. Key votes, by definition, are important to the parties in Congress, and possibly to the nation at large. It is difficult to imagine that a Representative would simply not find the time to vote for a key subject. Furthermore, given the importance of the vote and hence the strict party discipline enforced by party leadership, it is highly unlikely that a Representative that did really not have an opinion on a subject, did not vote, and thus risk disciplinary measures by his or her own party leadership, just because she did not have an opinion. It is much more likely, that Representatives did vote only present because their constituents wishes were against the party line, they however did not want to risk their party standing by voting against the party line.

To sum up, the modeling of not cast votes as missing values does not adequately reflect the nature of these votes. While it is clearly not possible to ascertain the motivation behind a not cast vote, discarding it entirely results in the discarding of potentially valuable information.

Obviously, the replacement of an unknown vote with a mean or median value is equally flawed, if party affiliation prediction is the goal. We therefore argue to include the not cast votes into the analysis, by changing the binary variable (with missings) to a ternary one (without missings). In terms of level of measurement this would yield a qualitative factor variable on a nominal scale with three factor levels: nay, yea and present. There, however, is more information included in this variable. Nay, yea and present are not only distinctive categories or types of votes, they also contribute to an outcome, that is to the passing or rejecting of a proposed law.

Either party that wishes to see a law passing Congress, will have to ensure that the majority of the valid votes (i.e. without the present votes) are yea votes. In that sense, yea votes not only present a category, but a category that is of greater value than a nay or present vote. This is to say that there is an inherent ordered structure in these values. We can thus elevate the level of measurement to ordinal scale.

DOI: 10.1051/itmconf/20171400009

Another interesting property of these variables is the actual weight of a vote. Clearly a yea vote is as supporting of the passing of a law as a nay vote is damaging. Since every vote is equally capable of determining the outcome, they posses the same weight. Hence, the distance between yea and nay votes on a line is the same. Similarly, present votes do not favor either side. They merely reduce the number of valid votes, and this benefits both sides struggling to further their respective causes.

So changing the interpretation of the variables from neutral voting behavior to supporting the motion and thus implementing a numeric coding of -1 for Nay, 0 for present, and 1 for Yea, allows an interval scale level interpretation of the voting variables.

This has profound consequences, as interval scaled variables provide access to a large number of statistical methods, that would have been out of the question otherwise. For the scope of this paper, we use the terms 2-value logic to refer to the coding of abstaining as a missing value. Using the information of abstention on the other hand leads to a 3-value logic model.

Another major flaw in the data set is the coding of party affiliation. While certainly Representatives of the 98th US Congress all belonged to either the Republicans or the Democrats, this only partly mirrors reality. Owing to a newly found strength and to thus resulting motivation, the Republicans managed to find some sort of coherence in their voting behavior in the 1980s. The Democrats, however, were made up of two quite distinct parties, that shared little more than a common name and party leadership. The case of the Democrats is widely recognized in political science literature [12, 13], yielding even CQA to break down democratic votes into Southerners and Northerners. The data set does not follow this tradition: Southern and Northern Democrats alike are coded as Democrats and Democrats only. However, we will not dive any deeper into this issue here, as the CQA data set is mainly of historic value, and we will here use a more up-to-date collection of votes.

2.2 Overfitting avoidance

Many machine learning tools excel at prediction. The here introduced method of RDTs falls in the so called class of supervised learners. This means that an algorithm is trained on predicting by showing it already predicted data. For instance, when the algorithm should be trained to recognize apples, it needs to shown different objects, some of them apples and some of any other type. It will then deduce the criteria that turn an apple into an apple. This then can be used to classify other objects where we do not yet know if it is an apple or not. The first set with identified objects is called the training set. The test set is then the collection of cases that are to be predicted.

However, when predicting, one must take care to strike a balance between a model that explains the data very well, but then cannot predict formerly unseen objects. For instance, the apple identification task might lead to a model that learned that every round object is an apple, and will therefore wrongly classify oranges as apples. To avoid this, it is best practice to repeat the training many times, to get an idea of how dependent the algorithm is on certain features. If there is fear that the algorithm wrongly overemphasizes certain features, an averaging or bagging approach like random forests can be used. These averaging approaches, however, are not a topic we will discuss here any further.

Instead, we focus on estimating the variance between different training runs. This is done by randomly splitting the data set into two parts: a training and a test set. These parts are then used for training and testing the prediction quality. This is repeated many times, with each time the case allocation to a training or testing role is varied randomly. This gives many prediction results, that then can be compared to arrive at an idea of the variance. If this variance is low, as it is in this case as we will see shortly, the no bagging is needed.

As we are dealing with roll call data here, it is worth noting that varying the random training-testing set approach is quite easy to implement. Another approach would be to use roll call data from a previous session. While this is in theory possible, it would be very difficult to accomplish in practice.

Table 1. Decision/Classification/Regression Trees as method in journal articles.

Journal	–2007	2008–2012
American Political Science Review	9	0
American Journal of Political Science	10	0
Public Opinion Quarterly	1	2

Table 2. W-NOMINATE scores as predictors in journal articles.

Journal	–2007	2008–2012
American Political Science Review	6	1
American Journal of Political Science	7	2
Public Opinion Quarterly	0	0

This would require that the roll call records of the testing period be somehow linked with those of the training period. Returning to the apple example: if the algorithm was trained on classifying objects because of their color and shape, but then is tested on the information of weight only, it will fail. Only if we could somehow deduce a link between color/shape and weight, could the algorithm be successfully applied. When resorting to the splitting of a single data set, these limitations do not apply.

2.3 Literature review

As noted above, the concept of regression and decision trees is not new. It's heyday in political science use, RDTs enjoyed right after their invention in the 1980s. In the following, we would like to give an overview of the literature relating to RDT applications in political science' three premier journals: American Political Science Review, American Journal of Political Science and Public Opinion Quarterly. These journals have been chosen as they have been continuously ranked as the journals with the highest impact factor in political science by Thomson Reuters Journal Performance Indicators.

The following Table 1 gives an overview of articles using RDTs as a method in those journals[1] Additionally, the use of W-NOMINATE scores as predictors in regressions was researched. These results[2] are found in Table 2. The periods have been broken down in two discrete areas, revolving around a split point in 2008. Especially with decision trees, there is a wealth of articles referring to qualitatively established decision trees, that are, however, unrelated to the here discussed methodology (e.g. [20]). Therefore, these articles have been removed.

As was demonstrated in this section, thought needs to be given to way voting data is being represented. These choices have serious consequences to the toolbox being available to modelers. As apparently the popularity of RDTs in social sciences has somewhat faded, we will now turn our attention towards the methods contrasted in this paper: regression and decision trees and W-NOMINATE based models.

3 Regression and Decision Trees

Regression and Decision Tree Analysis differs greatly from the formerly discussed regression analysis. While regression analysis starts with a statistical model, and tries to adjust its parameters to fit the

[1]The most recent articles using data driven RDTs are [14, 15]. [16] is one of the first texts of using data driven methods in social sciences, and takes a very critical stance.

[2]The most recent articles containing W-NOMINATE as content of regression models are [17–19].

model to the data, RDTA goes the same way backwards: its starting point is located within the dataset. From there the data is divided, or grouped along its variables, with the goal to create as pure groups as possible. To this end, the variables have to be rank ordered for their suitability of splitting and a split value has to be determined. Another problem with RDTs is that they can grow quite large. Therefore, a technique called pruning is employed, after the tree has grown, to eliminate all portions of the tree that are not important for the final result. Important here means contributing actively to the quality of the classification. For this some sort of purity measurement (i.e. how good does a node apply to only one of the groups) has to be developed.

RDTs can be used for a variety of tasks. One of the more common applications is the classification of data into different groups. [21] term decision trees that classify along a discrete feature classification trees, while trees that produce a clustering with variable membership rates are labeled regression trees. For the aim of this paper, we will combine both tree species under the term RDTs.

The central questions that need theoretical foundation in a RDTA "pertain to the criteria for splitting nodes and pruning the tree" [21, p. 270]. The former is a measure of divisibility as mentioned above, while the latter assess how complex a tree has become and how useful a node can be, purity. Splitting is rather straight forward, but unfortunately it is computationally unfeasible to try out all possible splits, and then prune the useless ones. So usually a greedy algorithm is used for growing the initial tree. For pruning that tree, a plethora of interestingness measures has been proposed in the literature. As one of the most important purity measures, and highly recommended by [21], the Gini index will be used in this paper. For the two classes case, the Gini index is defined as

$$GI = 2p(1 - p),$$

with p being the proportion in the second class. The lower the GI for a node, the purer it is, and the more information can be won from keeping the split (and possibly its successive branches) in the tree. Pruning is repeated for all of the nodes in the tree, and yields an optimal result. Of course, if the initial split got flawed by some means, pruning cannot help anymore. So due care must be payed by choosing the initial variable for the split. A possibility is, to not just randomly pick it, i.e. start the search at a random point, but to start from a variable that is rather pure itself.

Within R, the function rpart() provides the necessary functionality to grow and prune a tree. It provides different approaches in growing the tree that are suitable for classification tasks or for instance to group survival analysis data or to grow regression trees. The function is complemented by a plotting utility that produces dendograms of the grown trees. The default plotting capabilities are extended by [22].

A grown tree can then be used to predict the outcome for unseen cases. By examining a new cases values at the splitting nodes of the tree, the case is moved further down the tree until it finally arrives at a leave node that then assigns that case its predicted value.

4 W-NOMINATE Scores

An entirely different approach to the analysis of voting behavior is pursued by W-NOMINATE scores, as introduced by [23]. Here, a three-step multidimensional scaling method is applied iteratively to voting data. In essence, the idea behind W-NOMINATE is to reduce the many single vote variables into a few dimensions. So instead of describing a legislator's voting behavior in terms of hundreds or thousands of recorded votes, this behavior is condensed in some (i.e. less than 10, usually 2) newly created variables with the canonical interpretation of economic and lifestyle based divides. In that, this method is similar to principal component or exploratory factor analysis. W-NOMINATE and its extension, dynamic W-NOMINATE are de-facto standards for locating US electors in political space. Therefore, we refer the reader to established literature detailing the intricacies of W-NOMINATE.

Here, we will now focus on using W-NOMINATE scores to predict voting behavior for legislators whose intentions on a particular vote are not yet clear. To this end, the W-NOMINATE scores (w_1, $w_2, \ldots$) obtained from the past voting record are used as predictors in a logistic regression, modeling the voting behavior (Yea vs. Nay) for the vote in question:

$$ln\left(\frac{p_y}{1-p_y}\right) = \beta_0 + \sum_{i=1}^{5} \beta_i w_i$$

Here β_i denotes the regression coefficient for the i-th W-NOMINATE score. To ensure a properly fitting model, step-wise selection is being applied afterwards, to remove those terms that do not contribute to model quality.

This model can be used to arrive at a probability of voting yea for unknown legislators, given their past voting record and thus their W-NOMINATE scores. A predicted probability of mode than 0.5 is used to classify that legislator as a future yea-voter; otherwise the legislator will be assigned the class of nay-voter.

W-NOMINATE score estimation is implemented by the wnominate package in R [23]. This package, however, is based on FORTRAN code and takes a heavy toll computational wise. For instance, computing the W-NOMINATE scores on the 110th House in 5 dimensions took about 15 minutes on a reasonably sized computer. For comparison, the computation of a decision tree with the same input variables took mere 2 seconds.

5 Testing Methodology

To establish the suitability of RDTs for political science applications, we looked at the two distinct tasks. The first task centered on producing an analytically accessible model of voting behavior. The second task was to predict the outcome of key votes, using only prior voting behavior as predictors. We will discuss both tasks in turn now.

5.1 Analyzing voting behavior

A central virtue of any model is to not only fit the data it is based on well. Rather, a model algorithm also needs to produce a model that is transparent in that it allows researchers to interpret the modeling choices made, i.e. a black-box model is always of less value than one from which analytic insights can be deduced.

For the analytic scenario, the task was set to provide a model that allows a classification of a vote along arbitrary dimensions. This arbitrary dimension allowance was added for comparability of W-NOMINATE and RDTs. The idea is to be able to call a vote being like another vote or similar to political dimension. For the W-NOMINATE approach, this can be achieved by looking at the regression coefficients produced in the model described above. Since the W-NOMINATE scores are normalized to lie within a unit circle (or hyper sphere for higher dimensions), they can be compared with each other to establish which dimension has the largest influence on a vote.

Likewise, in RDTs the tree's variable selection can be interpreted as votes that are similar to the vote in question.

When assessing analytic model qualities, we looked for sensitivity to parametrization with regards to consistency. A different model specification should not lead to a completely different model.

5.2 Predicting voting behavior

For this task the following scenario was assumed. Before a key vote, individual representatives are contacted to inquire on their voting intentions. Additionally, legislators may have chosen to publicly state their intended behavior. Using this knowledge, and assuming that there are no last-minute mood swings in the legislators, the voting behavior of the remaining representatives is to be predicted. As additional predictors, this scenario permits the using of past voting behavior for every elector. Armed with this information, a decision tree can be grown following this formula:

$$V_j \sum_{i=1}^{j-1} V_i + P$$

Here, V_j is the vote to be predicted. The (possible) predictors for this vote is the past voting record for all votes (V_1 to V_{j-1}) and the representative's party membership.

Comparing the predicted voting behavior of the missing legislators to the the one actually observed later on, a quality measure for the prediction can be established: the proportion of the legislators' vote that was wrongly classified leads to the misclassification rate. Obviously, the lower the misclassification rate, the better or more accurate the prediction is.

A model that is able to successfully predict how legislators will vote, arguably captures the political process in detail. If that model is also able to be analyzed, i.e. is not a black box, scientists can gain insights on correlations between votes, i.e. examine what makes legislators vote the way they do. Further, any such model has useful applications in lobbying industries.

In the following, two models were pitted against each other. One model was formulated among more traditional lines: here, a logistic regression model used the W-NOMINATE scores of legislators to predict their voting behavior on a key vote. The W-NOMINATE scores were computed using all the votes in the data set that had occurred before the to-be-predicted vote was cast. The second model uses a decision tree to arrive at the same prediction. As input variables, the algorithm could choose from all previously cast votes.

The set of legislators was split randomly into known and unknown legislators. The assignment of each legislator followed a uniform random distribution. Different sizes of training sets were tried out and predictive performances compared. The set of known legislators was used to train, i.e. grow the decision tree or estimate the regression parameters. This tree/model was then used to predict how the unknown legislators would vote.

To establish the quality of a prediction, we examined how many legislators' votes were incorrectly predicted. Putting that number into relation with the total number of legislators to classify, the misclassification rate is computed. To ensure reliability of the obtained misclassification rate, a Monte Carlo technique, namely cross-validation, was used. This means that each classification was repeated many times over, splitting the data set randomly into known and unknown legislators.

A useful model, therefore, will be able to predict unknown legislators' voting behavior comparably accurate. Additionally, it should be able to do so with only few legislators being known. Also, the quality of the prediction should not be dependent on which legislators are known. This means, that not only the misclassification rate has to be low. The successful model also needs to exhibit a low level of variance between the cross-validation replications. A high level of variation would require many models to be computed. Combining these many models to arrive at a meta-model that allows also analytical insight is far from trivial. Low variation eradicates that need.

In this section we have stated two scenarios that can be used to answer questions in the field of political science. Further, indicators have been identified that allow judging a method's suitability. The analytical scenario requires to provide consistency even in the face of misspecification, while in the predictive scenario it is important to produce accurate predictions with a low variability. We will now apply both methods in both scenarios to data from the 110th US House, describing the obtained results.

6 Results

In the following we present the results using the aforementioned methodology on a specific set of voting data. The data set analyzed was that of the first session of the 110th US House of Representatives, as provided by [24]. This data set contains all the roll call votes of all Members at that time. The data set has 449 rows (legislators, some of them came into office or left office during the time covered) and 1865 columns (recorded roll call votes). Each field has one of 3 possible values, representing yea, nay and abstained. A number of votes have been identified as key votes by the editors of the Congressional Quarterly weekly reports. These votes were selected because they got wide media coverage, or controversial or particularly close. They can also be a test of presidential power. Key votes as identified by CQ have been used regularly in political science [25, 26]. To test the advantages of RDTs over the traditional W-NOMINATE-based regression, a random stretch of votes, namely Votes 200–250 were selected. Additionally, more in-depth analysis was conducted on a randomly selected key vote: 836.

6.1 Analysis

To gauge analytical model qualities, the entire data set of roll call votes was used to explain a singled-out key vote. To do that, both approaches were provided with either a 2-valued or 3-valued parametrization as detailed above. In the following, the estimated logistic regression parameters and effect plots for the resulting W-NOMINATE based models will be contrasted. After that, the regression and decision trees produced will be presented.

As stated above, a key value of any model is insensitivity to different ways of parametrization. One possibility to judge this quality is by looking at the models produced using a 2- and a 3-valued specification of voting behavior: does the consideration of abstainers alter the resulting model? When computing W-NOMINATE-based models, the step-wise feature selection algorithm is forced to remove the third component from 2-value model. As a result, the coefficients of both models vary strongly. One efficient way to render these differences are the effect plots that are given in Figure 1.

These effect plots detail, that both models lead to similar effects of most W-NOMINATE score dimensions. However, dimension three is removed from the 2-value model and dimension five has a much milder effect in the 2-value model than in its 3-valued brother. While the ultimate conclusion, that Vote 836 is a vote that can be explained well along the first W-NOMINATE dimension is not altered, the effect sizes between model specifications do vary considerably.

Tasking RDT-based models with producing analyzable models leads to dendograms of the respective trees. They are depicted in Figure 2. While both model parametrizations lead to similar choices in splitting points, the finer-grained model results in a more detailed tree. However, there are no contradictions.

6.2 Prediction

For the task of prediction, both approaches were asked to predict the voting behavior of a majority of legislators for a given vote. As source of information, the past voting behavior of all legislators and the voting behavior of a subset of legislators on the vote in question was provided.

The main quantity of interest here is the misclassification rate as defined above. First, the performance of both methods given a different number of known legislators as a training set was compared.

DOI: 10.1051/itmconf/20171400009

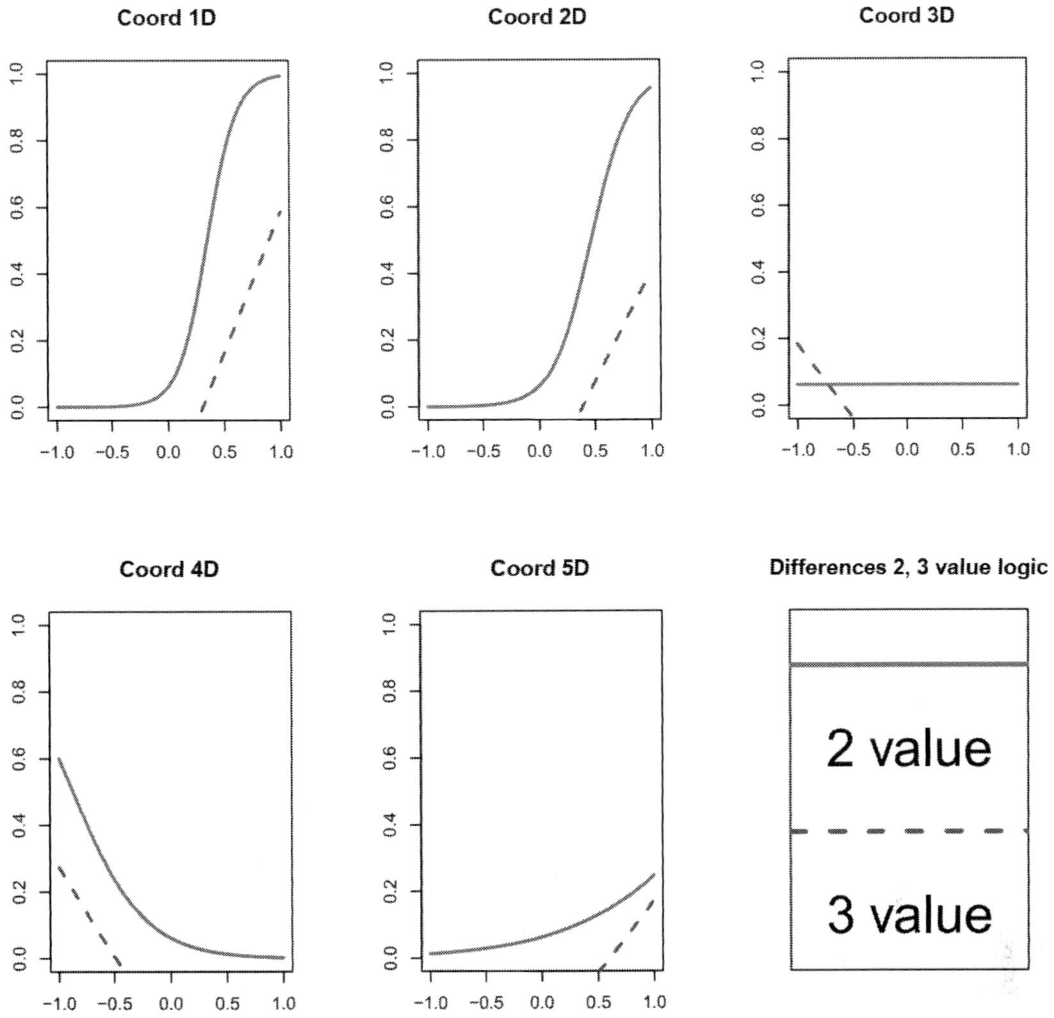

Figure 1. Comparing the effects of the respective W-NOMINATE-based models parametrized with 2- and 3-value logic respectively. If the choice of parametrization is irrelevant, the effect curves should be identical.

These comparisons were done on a randomly selected key vote. The procedure of randomly selecting a given share of legislators as training set was repeated 250 times to produce results indicative of variability. The performances were compared using one-sided two-sample t-tests with the null hypothesis that WNR-based mean misclassification rates are lower (or equal) than those produced by RDTs. The level of significance has been adapted for multiple testing using Bonferroni correction. These results are shown in Figures 3 and 4.

It is striking to see the performance differences between the two approaches. While WNR produces a constant quality so to speak, RDTs improve vastly, once a threshold number of legislators used in the prediction is exceeded. What's more, is also that RDTs exhibit a drastically lower vari-

DOI: 10.1051/itmconf/20171400009

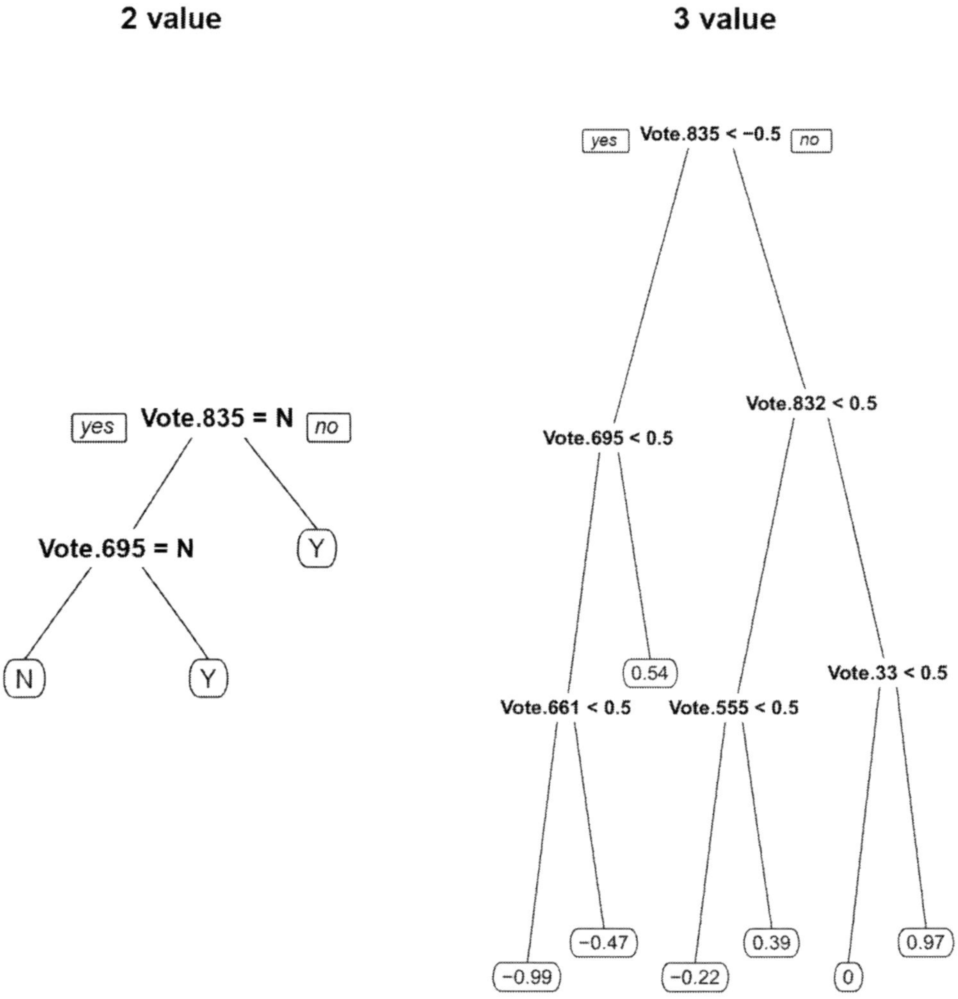

Figure 2. Contrasting decision trees grown from 2-value and 3-value logic models.

ance across replications. This behavior of low variance and thresholds is also present in all the other key votes, as demonstrated in Figure 5. Note that, due to WNR's computational expense, no computations for all key votes were carried out for the WNR method.

Albeit key votes are arguably the most interesting ones to predict, both methods were also tested using a randomly selected stretch of consecutive roll call votes. The task was again to predict the voting behavior of unknown legislators using only prior voting record as predictors. Each computation was repeated 25 times using different randomly generated training and testing set splits. A series of t-tests was computed to test the alternative hypothesis that RDT's misclassification rate was lower than the one generated by W-NOMINATE based prediction models. The p-values of the resulting tests are

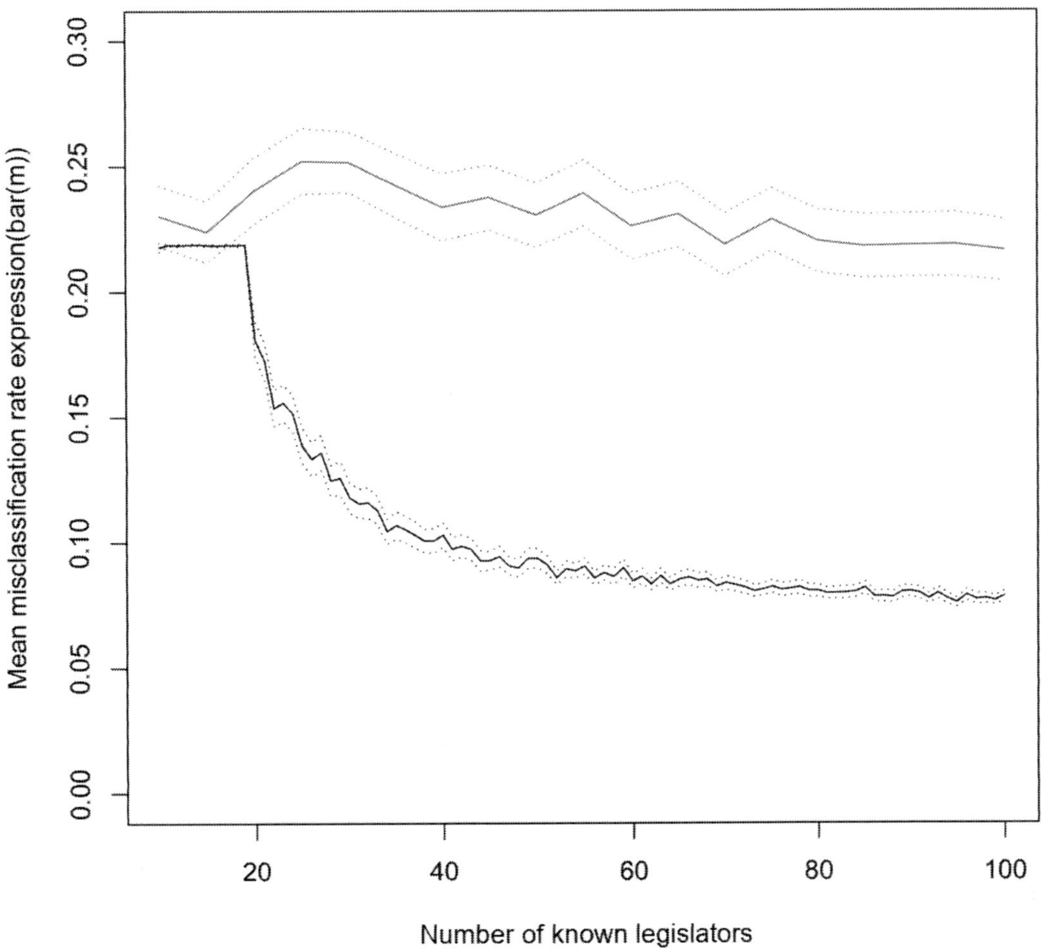

Figure 3. Comparing prediction performance for WNR and RDTs across different training set sizes for key vote 836. The bottom line is the mean misclassification rate of RDTs, the top line that of WNR-based prediction models. The shaded area are 95% confidence bands. Mean and variance has been established through cross-validation random sub-sampling with 250 replications.

given in Table 3. When applying a Bonferroni corrected level of significance of 0.001, 12 votes were predicted significantly more accurately using RDTs than using W-NOMINATE.

7 Discussion

When comparing methods used for an analytic purpose, the consistency of the model across (inadvertently) chosen parametrizations is a striking factor. The effects resulting from different modeling

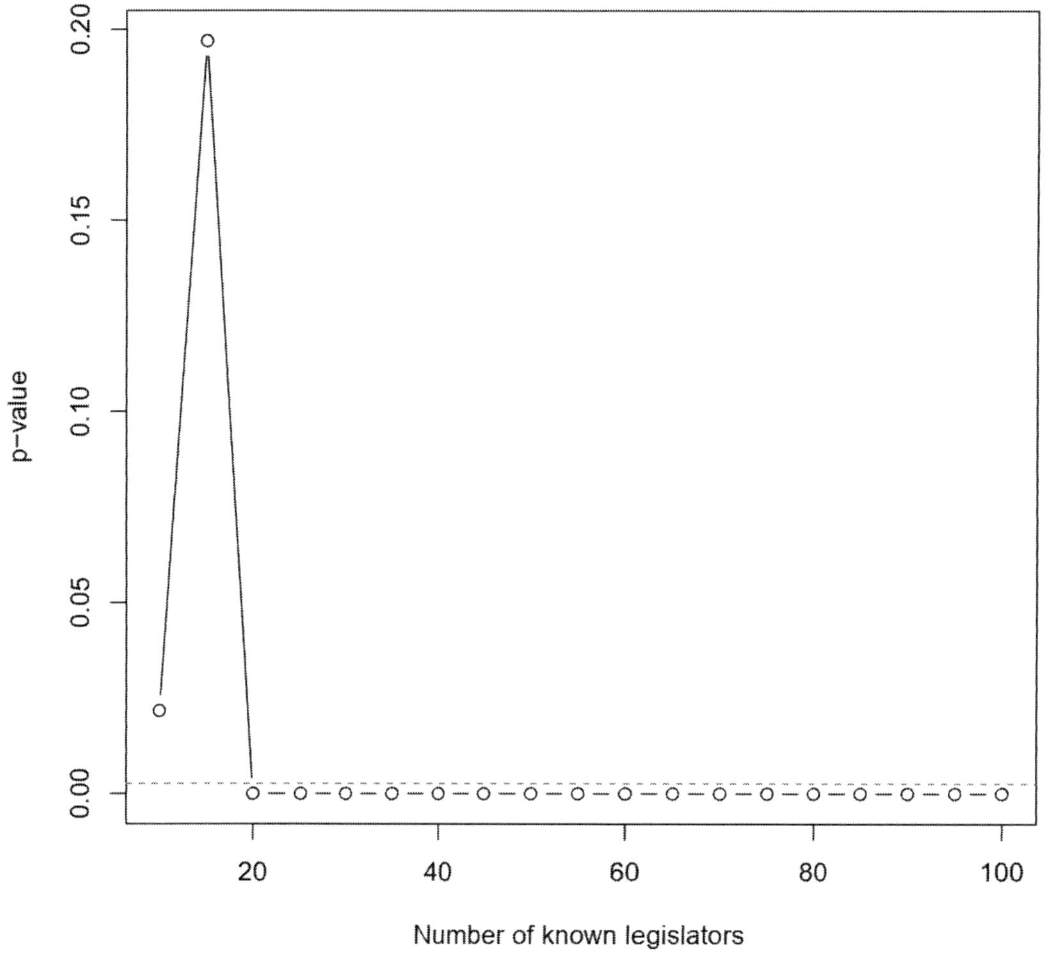

Figure 4. P-values for one-sided two-sample t-tests, testing the alternative hypothesis that $\bar{m}_{RDT} < \bar{m}_{WNR}$ for different numbers of known legislators. The dashed horizontal line indicates the Bonferroni corrected level of significance (0.0026316). Every p-value below that line indicates a significantly better performance of RDTs compared to WNR-based models.

choices in the W-NOMINATE based models, indicate that consistency here is limited. Especially in the higher, and arguably less important dimensions, the observed effects between a 2-value and a 3-value logic differ. Especially noteworthy is the omission of the third dimension from the model specified by a 2-value logic.

The RDTs are somewhat similar in this respect. While the 3-valued model results in a superset of the votes selected to describe the 2-value model, there are no contradictions. However, the larger tree allows for finer grained classifications.

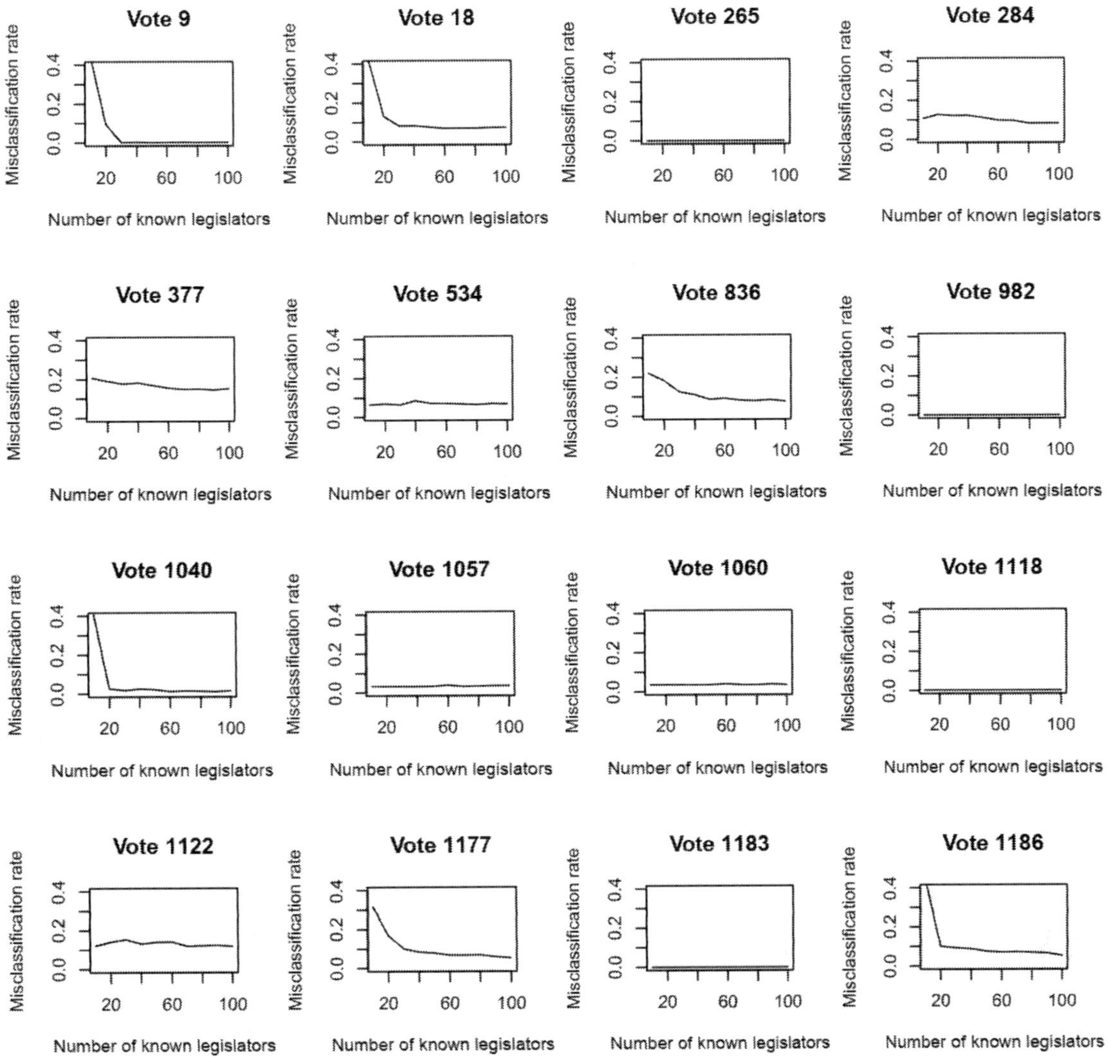

Figure 5. Performance of RDTs in relation to the number of known legislators the prediction is being based on for all key votes. Horizontal lines indicate votes that were too easy to predict, i.e. most legislators voted in the same way. Computations averaged over 25 replications.

To summarize, for analytic purposes, W-NOMINATE based and RDT based models are on par with each other. Given the computational expense required by W-NOMINATE score estimation, however, RDTs appear to be more suited for data intensive analysis.

Turning to predictions, the differences between both methods become more evident. RDTs exhibit consistently a lower variance over cross-validation replications. RDTs apparently require a threshold level of known legislators to be grown from: only then can they deliver accurate predictions of voting behavior. However, this threshold number is rather low being in the range of 15 to 20. Once this threshold is reached, RDTs consistently produce significantly better results than W-NOMINATE

Vote	p-Value	Vote	p-Value	Vote	p-Value	Vote	p-Value	Vote	p-Value
201	0.183	211	1	221	0.002	231	0.001	241	1
202	1	212	0.126	222		232	0.001	242	0.003
203	0.197	213	0.001	223	0.001	233	0	243	0.059
204	0	214		224		234		244	0.561
205	0	215		225	0.536	235		245	0.001
206	0.975	216		226		236	0.349	246	0.023
207	0.004	217	0	227	1	237	0.036	247	0.008
208	0	218	0	228	0.012	238	1	248	0.121
209	0	219	0.001	229		239	0.058	249	0.013
210	0	220	0.005	230	0.071	240	0.139	250	0.004

Table 3. P-values for one-sided two-sample t-tests, testing the alternative hypothesis that $\bar{m}_{RDT} < \bar{m}_{WNR}$ for a stretch of 50 votes. Missing entries result from W-NOMINATE based models not converging.

based models. This observation holds for key votes. Ordinary, less important votes, however, are somewhat different. Here RDTs still outperform W-NOMINATE based models most of time. On the other hand there are cases where W-NOMINATE delivers better predictions.

We conclude from this, that RDTs have certain advantages over W-NOMINATE based models. For one part, they are much faster to compute. Given the wealth of voting data that is now available, this is an advantage that should not be missed and is likely to grow in importance as big data is being gradually introduced to political science.

Another key observation is, that they produce transparent models that can be used to gain analytic insights in political processes. While W-NOMINATE also has much wider general applications, W-NOMINATE based regression models are not terribly useful in understanding the relationships of votes. RDTs on the other hand are easily interpreted. Finally, prediction accuracy of RDTs is extremely high and almost always better than that of W-NOMINATE based models.

While both methods have their advantages and disadvantages, we find that RDTs are capable tools that should be used more often in the political science community.

References

[1] A. Asuncion, D.J. Newman, *UCI machine learning repository* (2010), "http://archive.ics.uci.edu/ml"

[2] K.W. Bauer, S.G. Alsing, K.A. Greene, Neurocomputing **31**, 29 (2000)

[3] E.H. Han, G. Karypis, V. Kumar, B. Mobasher, Bull. Tech. Committee on Data Eng. **21**, 15 (1998)

[4] E.R. Hruschka, E.R. Hruschka, N.F.F. Ebecken, *Feature Selection by Bayesian Networks*, in *Advances in Artificial Intelligence*, edited by A.Y. Tawfik, S.C. Goodwin (Springer, 2004), Vol. 3060 of *Lecture Notes in Artificial Intelligence*, pp. 370–379

[5] E.R. Hruschka, E.R. Hruschka, N.F.F. Ebecken, *Missing Values Imputation for a Clustering Genetic Algorithm*, in *Advances in Natural Computation*, edited by L. Wang, K. Chen, Y.S. Ong (Springer, 2005), Vol. 3612 of *Lecture Notes in Computer Science*, pp. 245–254

[6] W. Davies, P. Edwards, *Distributed Learning: An Agent-Based Approach to Data-Mining*, in *Machine Learning-95*, edited by D. Gordon (AAAI Press, 1995)

[7] M.A. Potter, Ph.D. thesis, George Mason University, Fairfax, Virgina (1997)

[8] J.E. Jackson, The American Political Science Review **65**, 451 (1971)
[9] J. Clinton, S. Jackman, D. Rivers, American Political Science Review **98**, 355 (2004)
[10] M. Cohn, ed., *Congressional Quarterly Almanac 1984* (Congressional Quarterly, Washington, D. C., 1985)
[11] G.L. Hager, J.C. Talbert, Legislative Studies Quarterly **25**, 75 (2000)
[12] R. Fleisher, The Journal of Politics **55**, 327 (1993)
[13] M. Thomas, American Journal of Political Science **29**, 96 (1985)
[14] D.P. Green, H.L. Kern, Public opinion quarterly **76**, 491 (2012)
[15] G.R. Murray, C. Riley, A. Scime, Public Opinion Quarterly **73**, 159 (2009)
[16] H.J. Einhorn, Public Opinion Quarterly **36**, 367 (1972)
[17] J. Richman, American Political Science Review **105**, 151 (2011)
[18] J.W. Patty, American Journal of Political Science **52**, 636 (2008)
[19] E.G. Juenke, R.R. Preuhs, American Journal of Political Science **56**, 705 (2012)
[20] D. Skarbek, American Political Science Review **105**, 702 (2011)
[21] T. Hastie, R. Tibshirani, J. Friedman, *The Elements of Statistical Learning*, Springer Series in Statistics (Springer, New York, 2001)
[22] S. Milborrow, *rpart.plot: plot rpart models. An enhanced version of plot.rpart* (2011), `http://CRAN.R-project.org/package=rpart.plot`
[23] K. Poole, J. Lewis, J. Lo, R. Carroll, Journal of Statistical Software **42**, 1 (2011)
[24] K. Poole, J. Lewis, *110th house roll call data*, Published online (2010), `http://www.voteview.com/house110.htm`
[25] S.A. Shull, J.M. Vanderleeuw, Legislative Studies Quarterly pp. 573–582 (1987)
[26] J.N. Victor, N. Ringe, *Legislative Caucuses as Social Networks in the 110th US House of Representatives*, in *Networks in Political Science Conference, Cambridge, MA* (2008)

AUTHOR INDEX